Basic metric surveying.
Third edition

D0841413

Basic metric surveying

Third edition

W S Whyte
Principal Lecturer,
School of Land and Building Studies,
Leicester Polytechnic

and

R E Paul
Senior Lecturer,
Leicester Polytechnic

Butterworths
London Boston Durban Singapore Sydney Toronto Wellington

First published 1985

© Butterworth & Co. Publishers Ltd 1985

British Library Cataloguing in Publication Data

Whyte, W. S.
   Basic metric surveying.——3rd ed.
   1. Surveying
   I. Title   II. Paul, R. E.
   526.9′024624   TA549

   ISBN 0-408-01354-0

Library of Congress Cataloging in Publication Data

Whyte, W. S. (Walter S.)
   Basic metric surveying.

   Bibliography: p.
   Includes index.
   1. Surveying.   2. Metric system.
   I. Paul, R. E.   II. Title.
   TA545.W69   1985   526.9′024624   85-3819

   ISBN 0-408-01354-0

Photoset by Butterworths Litho Preparation Department
Printed by Anchor-Brendon Ltd., Tiptree, Essex

Contents

Preface to the third edition

This edition has been completely re-written to provide a modern approach to surveying which will take account of the developments which have taken place over the last few years. The primary aim remains, however, as the provision of a guide to current practice for non-specialist surveyors in the various professions involved in the construction industry.

Obsolete methods such as plane tabling and the use of logarithms have been omitted. The sections on optical distance measurement methods and equipment have been reduced and replaced by a consideration of electromagnetic distance measurement methods and equipment. The detail on instruments generally has been reduced.

All calculations are now assumed to be made using an electronic calculator or a microcomputer and examples of the use of surveying software are provided in the relevant sections.

The traditional survey methods are covered, but an emphasis is placed on the methods of bearing and distance survey which are more widely used since the advent of electronic measurement and computation equipment.

The content should be generally suitable for students preparing for degrees and diplomas in architecture, building, building surveying, quantity surveying, estate management and town planning. It may also be of value to engineers who are not specializing in engineering survey.

We are indebted to our colleague, Peter Swallow, Principal Lecturer in Building Performance at Leicester Polytechnic, for the new chapter on Site Investigation.

WSW
REP
1985

Part 1

Introduction

Chapter 1

Introduction to surveying

1.1 Surveying

Surveying operations are very often associated with the preparation of *survey drawings*, which may be maps, plans, sections, cross-sections or elevation drawings. The survey of land, together with the natural or man-made features on or adjacent to the surface of the Earth, usually requires two-dimensional drawings of three-dimensional objects. Such drawings are usually produced on either a horizontal or vertical plane.

A drawing on a horizontal plane is known as a plan or map. A plan is a true-to-scale representation, while a map may contain features which are represented by conventional signs or generalized symbols, these not being true scale representations.

For example, on a map a feature might be represented by a dot, roads may be drawn to a standard width, or a building with many juts and recesses may be drawn as if it were a simple rectangle on plan.

Drawings in a vertical plane are known as sections, cross-sections or elevations. An elevation is a side or end view of an object, such as a view of the end of a building. A section is a vertical 'plan' of a line through a building, or a line of a proposed road, etc., as it would appear upon an upright plane cutting through it. A long section such as along a proposed road or rail route is known as a longitudinal section, whilst sections taken at right angles to the longitudinal line are known as cross-sections.

1.2 Objects of surveying

Surveying techniques may be considered to be used for three distinct purposes as follows.

1.2.1 Surveying for the preparation of maps, plans, etc.

The determination of the relative positions of natural and artificial features on, or adjacent to, the surface of the Earth, so that they may be correctly represented on maps, plans or sections.

1.2.2   Setting out

The setting out on the ground of the positions of proposed construction or engineering works. The information on the new works is normally found in setting out documents which usually include some of the drawings described above.

1.2.3   Computations such as areas or volumes

The execution of calculations for land areas, for earthworks volumes, etc., either based on 'field' measurements or on measurments abstracted from maps, plans and sections.

1.3   Classifications of surveying

Over the years, surveying has been classified or sub-divided in a variety of ways.

1.3.1   Plane and geodetic surveying

When the features of the ground are shown on a map or plan, they are shown on a flat sheet of paper or other medium, which represents a horizontal plane. Since a horizontal plane touching the Earth's surface at one point will be tangential to the Earth at that point, it is evident that the Earth's surface cannot be accurately represented on a plane. In ordinary geometry, the angles of a triangle always add up to 180°, but on the surface of a sphere the angles of a triangle add up to more than 180°. The Earth, fortunately, is such a large spheroid that for surveys of limited extent there will be no appreciable difference between measurements assumed to be on a plane surface and measurements made on the assumption of a spherical surface.

Where measurements cover such a large part of the Earth's surface that the curvature cannot be ignored, then the operations are termed *geodetic surveying*. Where surveys cover such a small part of the Earth's surface that curvature can be ignored, then the operations are termed *plane surveying*.

The area which can be regarded as a plane will depend upon the accuracy required of the survey, but may be taken as up to $300\,\text{km}^2$ or more. The types of survey dealt with in this book are all aimed at tasks well within this arbitrary limit.

1.3.2   Classification according to purpose or use

1.3.2.1   *Geodetic survey*

A survey of great accuracy which not only takes into account the curvature of the earth but also provides control for surveys of lower accuracy.

1.3.2.2   Topographical survey

A survey which results in the production of maps and plans showing *topography*, i.e. the natural and man-made features on the surface of the earth.

1.3.2.3   Cadastral survey

A survey for the preparation of plans showing and defining legal property boundaries.

1.3.2.4   Engineering survey

A survey preparatory to, or in conjunction with, the execution of engineering works such as roads, railways, dams, tunnels, sewage works and construction works generally.

1.3.2.5   Mining survey

1.3.2.6 Hydrographic and other surveys.

1.3.3 Classification according to the equipment or techniques used

(1) Chain survey
(2) Traverse survey
(3) Tacheometric survey
(4) Photogrammetric survey
(5) Plane table survey, etc.

Some of these, and others, will be met later in the text.

1.4   Survey tasks

Dependent upon the object of the particular survey, the surveyor will be required to carry out one or more of the following tasks.

1.4.1   Detail surveys

Surveys for the supply (i.e. the survey or measurement) of detail. *Detail* is the name given to the natural and man-made features of the survey area. Generally, detail is taken to exclude *relief* (heights).

1.4.2   Heighting

Heighting denotes the supply of height measurements. *Height* is the vertical distance of a feature above or below a datum or reference surface.

1.4.3   Control

A large survey site requires many sets of measurements and calls for methods of tying these measurements together so as to produce an accurate survey. These methods are known as *control*.

1.5   Scale and scales

Reference has already been made to *scale* and in the next section the word is used again but with a different meaning. It is essential, at this stage, that the reader understands clearly what is meant by the word 'scale' with respect to survey.

Maps, plans, etc., are all proportional representations on paper or other drawing material of actual features on the ground. The ratio of a dimension on a map, plan or other drawing to the same dimension on the ground is known as the *scale* of the map, plan, etc.

Scale may be indicated in several ways, e.g. by a statement that 1 mm on the paper represents 0.5 m on the ground (often abbreviated to 1 mm represents 0.5 m, or 1 mm represents 500 mm). This ratio of 1 to 500, traditionally written 1:500 or 1/500, is termed the *representative fraction*. It may be abbreviated to RF.

Again, the scale of a map or plan may be indicated by a *scale line* (or *linear scale*) as described in Chapter 4. Scales vary enormously – the International Map of the World is at a scale of 1:1 000 000, town maps may be at 1:10 000 scale, some cadastral plans are at a scale of 1:1250, and occasionally component and assembly drawings may be as large as 1:1.

Scales may be said to be large or small (sometimes medium) but there is no definite dividing point. The Ordnance Survey (O.S.) has a dividing point between the 1:50 000 and 1:25 000 scale maps. 1:50 000 scale and smaller are small-scale maps while 1:25 000 and larger are large-scale maps. One scale may be said to be larger than another, on the basis that the larger the denominator the smaller the scale. A user may obtain distances from a map with the aid of a pair of dividers and the scale line. The surveyor can do likewise but the surveyor may also have to produce the map or plan from the survey measurements. To enable these tasks to be carried out to an acceptable accuracy, the surveyor makes use of a piece of draughting (drawing) equipment known as a *scale* or *scale rule*. The scale looks similar in shape and size to a rule.

1.6   Basic survey methods

The following sections outline the basic survey methods used in the supply of detail, heighting and control.

1.6.1   Methods of supplying detail

In *Figure 1.1*, imagine that the line AB represents a straight hedge, and point C represents a tree at some distance from the hedge. If it is required

Tree

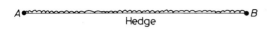

Hedge

Plan

Figure 1.1

to draw a plan of the area, then the hedge can be represented by a single line drawn to the appropriate scale length on the paper. How can the position of the tree be located and plotted on the plan? A variety of methods are available, as in the following sections.

1.6.1.1   Offsets (rectangular co-ordinates)

Measure the perpendicular distance (*offset*) from the tree to the hedge at point D (*Figure 1.2(a)*), and the distance from one end of the hedge, say A, to point D. On the paper, scale off distance AD and mark D. Set up a right angle at D, and set out the scale length of DC to locate C.

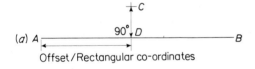

Offset/Rectangular co-ordinates

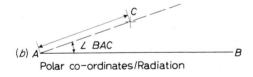

Polar co-ordinates/Radiation

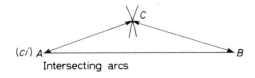

Intersecting arcs

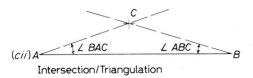

Intersection/Triangulation

Figure 1.2

Offsets are short measurements at right angles to a measured straight line. This measured straight line is often known as a *chain line*, hence the method of survey is generally known as *chain survey* and the procedures involved are known as *chaining*. The method is described in Chapter 4.

1.6.1.2   Bearing and distance (polar co-ordinates)

Measure the horizontal angle BAC (a *bearing*) on the ground and measure the distance from hedge end A to the tree. On the paper, use a protractor to set out the angle BAC, then set out the scale length of AC to locate C. The method is described in Chapter 14.

1.6.1.3   Intersection

A point of detail may be located by the intersection of a minimum of two lines, one from either end of a base of known length. The lengths of the intersecting lines can be measured from either end of the base line (see (i) below) or alternatively the bearings relative to the base line could be measured as in (ii).

(i) Measure the distances AC and BC on the ground. On the paper, using compasses, swing an arc from A with radius equal in scale length to AC, and a similar arc from B, with radius equal in scale length to BC. The intersection of the arcs locates the point C. This simple method is often used to measure to points of detail in a small area, single-handed. It is also used in conjunction with chain survey.

(ii) Measure the horizontal angles BAC and ABC on the ground. On the paper, set out these angles by protractor, and their intersection locates point C.

All four methods, shown in *Figure 1.2*, may be used for setting out detail and in map revision. (See Chapters 16 and 17.) The choice of method, or combination of methods, will depend upon the size and shape of the land, how it is currently used, whether or not heighting and/or control are required, the staff, equipment and time available, and usually also the cost.

1.6.2   Methods of supplying height (spot heights, contours, gradients)

1.6.2.1   Levelling

The commonest method of supplying heights on a site plan (*spot heights* and *contours*) or in setting out (including gradients) is by the use of a *surveyor's level* and a *levelling staff*. The staff, which might be described as a giant rule, is available in a variety of lengths. In use, the staff is held vertically by an assistant and read by the surveyor through a *level* which is basically a tripod-mounted telescope which may be set up in such a way that all lines of sight through the telescope are horizontal (*Figures 1.3* and *1.4*).

It will be observed from *Figure 1.3* that the telescope height is 25 + 1.3 = 26.3 m. The surveyor now rotates the telescope in the horizontal plane and an assistant moves the staff on to a peg (*Figure 1.4*).

If the second staff reading is 0.9 m, then the height to the top of the peg is 0.9 m below the horizontal line of sight, that is, 26.3 − 0.9 = 25.4 m.

There are limitations on the length of sight which may be used, so if it is necessary to transfer heights over a large horizontal distance then the operation has to be repeated. The method is described in Chapters 8 and 9.

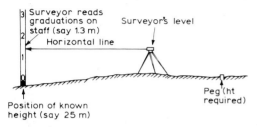

Figure 1.3

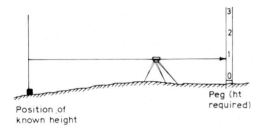

Figure 1.4

1.6.2.2   Trigonometric heighting

This method, an alternative to levelling, is not generally as accurate, but it may be useful when the ground is undulating or steep. Instead of a surveyor's level, a *theodolite* is used to read angles of elevation or depression; then provided the horizontal or slope distance is known, the difference in height can be calculated by elementary trigonometry. The method and use are described in Chapter 14.

1.6.2.3   Barometric heighting

Atmospheric pressure varies with height, thus a *barometer* which records atmospheric pressure may be used to obtain height measurements. Instead of being calibrated in inches, millimetres of mercury or millibars, it can be calibrated in metres, then the instrument is known as an *altimeter*. Anyone who has studied a barometer or a weather map will appreciate that atmospheric pressure changes continually without any change in height, hence the method is often not of sufficient accuracy. However, the method is frequently used in many parts of the world for heighting points for small-scale maps. It is not described further in this book.

1.6.2.4 Hydrostatic (water) levelling.

One of the oldest forms of levelling is to use a glass U-tube partly filled with water, then sight along the horizontal line (imaginary) joining the tops of the two water surfaces. Today the glass tube has been replaced by a flexible one with glass or transparent plastic gauges at each end.

Building site operatives use a flexible tube of approximately 30 m in length, but tubes of up to 20 km in length have been used in geodetic surveys. It is essential to note that all air bubbles must be removed from the tube, and this is difficult over long lengths. Another method, often overlooked, is to use the natural level of a body of water, provided that the surface is calm and the water flow is negligible. See the Bibliography for further reading.

1.6.3   Methods of supplying control

1.6.3.1   Use of a base or base line

A *base* or *base line* is simply a straight line of known length extending through, adjacent to, or within the area of the survey. Its length should preferably approximate to the maximum length of the site, but this depends upon the site conditions and the methods used to supply detail. The base controls the scale of the survey, that is to say it ensures that the completed survey is neither too small nor too large. Hence the base must be measured with great care, using carefully checked measuring equipment. The base may also serve to prevent distortion, that is to say to ensure that the completed survey is true to shape. Both the methods of supplying detail by intersection make use of a base, and the base principle is used with the other methods of supplying control.

1.6.3.2   Triangulation

Triangulation means a network of triangles tied to a measured base line, all the angles of the triangles having been measured. The method was used by the O.S. to supply the geodetic control for the mapping of Great Britain, and it involved thousands of triangles, an immense computational feat in those early days. As a means of geodetic control it is now obsolete.

The method may, however, be useful in conjunction with traversing, using a small number of triangles, to assist in maintaining accuracy. (See Section 1.6.3.4 below and Chapter 6.)

1.6.3.3   Trilateration

This again means a network of triangles, but in this case all the side lengths of the triangles are measured and not the angles. The method is the basis of all chain surveying techniques, whether used for the surveys of land, building plots, or actual building plans. It is relatively simple to plot with the aid of compasses, scale and a pencil.

Note that this technique is only known as *trilateration* when used in higher order work and the control observations have to be computed. If the network is part computed and part plotted then it is known as a *semi-graphic solution.* The higher order work is little used in surveys such as are covered by this text, but again on a few occasions it may be useful in conjunction with traversing.

1.6.3.4   Traversing

This is the most favoured and most flexible method of providing control for surveys, particularly in urban areas where the construction of triangles is difficult and sometimes impossible.

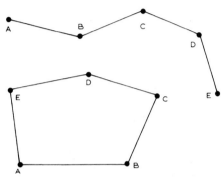

Figure 1.5

A *traverse* consists of a series of straight lines whose lengths and bearings can be determined (*Figure 1.5*). The lines are known as *legs* and the end-points as *stations*. The legs may be used either as chain lines for detail supply or for holding a network of chain lines. Alternatively, detail may be supplied by polar co-ordinates from the traverse stations (detail by bearing and distance). Traverses may also commence from and close at either end of a base line. (See Chapter 6.)

1.7   Principles of survey

1.7.1   Control

Each survey should be provided with an accurate framework, the lower order work (detail and heighting) being fitted and adjusted to the framework. The traditional expression is *work from the whole to the part.*

1.7.2   Economy of accuracy

The standard of accuracy aimed at should be appropriate to the needs of the particular task. As a general rule, the higher the standards of accuracy then the higher the cost in time and money.

1.7.3   Consistency

The relative standards of accuracy of the various classes of work (control, detail, heighting) should be consistent throughout a survey task.

1.7.4 Independent check

It is desirable that every survey operation (fieldwork, computations, plotting) should either be self-checking or be provided with an independent check.

1.7.5 Revision

Surveys should, where possible, be planned in such a way that their later revision or extension may be carried out without the necessity of having to carry out a complete new survey.

1.7.6 Safeguarding

The results of survey work (survey markers, field and office documents) should, so far as possible, be kept intact for possible use at a later date, e.g. in a revision survey.

1.8 Accuracy, precision and error

1.8.1 Accuracy

Since no measurement is ever completely accurate, surveys are usually described as being to a certain standard of accuracy. *Accuracy* may be defined as the conformity of a measurement to its true value. The accuracy of a measurment is often quoted as a representative fraction, that is to say as the ratio of the magnitude of the error to the magnitude of the measured quantity. *Error*, here, means the difference between the true value and the measured value of a quantity. If, in measuring a line 1 km long, the expected total error is ± 0.01 m, then the error is 1 part in 100 000. (0.01 m/1000 m $= 1/100\,000$). This is an extremely high accuracy, not often attained in plane surveying. On the other hand, if an error of 0.01 m were made in measuring a line 2 m in length, then the standard of accuracy is 0.01 m in 2.0 m, or 1/200, and this is usually too low for any type of work. The greater the degree of accuracy required in a survey, the greater will be the expenditure in effort, time and money. The standard of accuracy must be in keeping with the size and purpose of the survey.

1.8.2 Precision

Precision denotes the degree of agreement between several measures of a quantity. If a quantity is measured several times, then the degree of agreement between the measures is the precision of that set of measures. It must be noted that a high degree of precision does not indicate great accuracy. The classic example is the darts player who, while aiming for the 'bull', places all three darts in 'double one' – the player is throwing with some precision (all darts close together) but the result is not very accurate (should have landed in the bull).

1.9   Types of error

It should be understood that no measure in a survey is ever exact – every measurement, whether linear or angular, contains errors. The types of error must be appreciated, and their relative importance and care taken to keep them to a minimum appropriate to the task in hand.

1.9.1   Gross errors, or mistakes, or blunders

These are due to carelessness by the surveyor, such as reading a level staff as 2.415 m instead of 3.415 m, or noting a distance measured with the chain as 5.45 m instead of 15.45 m. These can only be eliminated by the use of suitable methods of observing, booking, computing and plotting designed to show them up.

1.9.2   Systematic and constant errors

These are errors which always recur in the same instrument or operation, and they are cumulative, that is to say their effect will increase throughout the survey. As an example, if a nominal 20 m chain has been stretched (by hard usage) by an amount of 0.05 m, then every time it is laid upon the ground there will be a constant error in distance measurement of 0.05 m. If, in measuring the length of a line, the chain is laid down ten times, the length of the line will be noted as 200 m, but the true length will be 200.5 m. The error will have accumulated to $10 \times 0.05\,\text{m} = 0.5\,\text{m}$. This error would be regarded as being negative, since its effect is to make the measured length appear less than the true length. Conversely, if a chain is shortened due to bent links, etc., then it will cause cumulative positive error, since it will measure the line as being longer than it actually is. Systematic errors are of varying magnitude, such as those due to tape malalignment.

These errors may be guarded against by using suitable operational methods, by standardizing equipment, and by applying appropriate corrections to the actual measurements.

1.9.3   Random or accidental errors

Errors which remain after the errors described above have been eliminated or reduced are known as *random* or *accidental errors*. These are small, they tend to compensate one another, some are positive and some are negative and they are not cumulative in the general sense. They may be due to small imperfections in equipment, to changing environmental conditions, or to minor discrepancies such as the surveyor possessing a personal bias like a tendency always either to underestimate or overestimate the value of some reading.

In practice, these small errors and some cumulative errors will remain on completion of the measurements, and in a control network, in plotting and in setting out they must be distributed so as to minimize their effect on the final result.

1.10   The use of microcomputers

Large mainframe computers have been used for surveying since the early days of computing, usually in batch mode for tasks which entailed handling large volumes of data and high speed 'number crunching'. Such machines are also used with high-speed plotters for automatic draughting of maps and plans, in national survey organizations and specialist surveying companies. These computer applications, however, are not considered here, since we are concerned with the 'occasional surveyor' with limited resources.

Surveying calculations were formerly carried out using tables of logarithms or tables of trigonometrical functions with a mechanical calculator. These methods were superseded by the electronic calculator which, in turn, is being replaced by the microcomputer, some models of which are already small enough to be used in the field. Today every student should have access to a microcomputer, and they are being used increasingly in professional offices.

Like the calculator, the user need have no knowledge of how the microcomputer works, but they must have access to suitable software if they do not wish to spend a great deal of time learning how to program. Provided the software is robust, easy to use interactively and designed to perform the various survey tasks effectively, the microcomputer may be regarded simply as another tool for the surveyor's use.

The present-day use of the microcomputer as a basic surveying tool is illustrated here by reference to the MICROSURVEY Surveying Systems software. In order that the theory of each topic may be covered in the usual way, the microcomputer applications are kept separate and dealt with at the end of each relevant chapter, with example printouts in most cases.

1.10.1   MICROSURVEY Surveying Systems

All the applications illustrated refer to MICROSURVEY Version 2.1, which was originally developed to run in 32 k of memory and is 'officially approved' by Commodore for all their machines. Version 2.1 is also available on 8-bit and 16-bit machines running under the CP/M-80 and CP/M-86 or MS-DOS operating systems respectively.

A system consists of 15 interactive programs on a single floppy disk, the first program on the disk only being loaded by the user. This is a System Menu program, from which the user selects each program as required, the individual application program being automatically loaded by the system. All programs are 'menu-driven', and when a program run is terminated control returns to the System Menu.

Programs are designed for individual applications, but they are integrated where practicable. Thus, most programs can create data files on disk, and some programs can read files created by other programs, e.g. the Traverse program can read in linear and angular data from files written to disk by the Lines and Bearings programs respectively.

In some computer systems the storage capacity of a single floppy disk is insufficient to hold a complete Version 2.1 system. In these cases, sub-systems may be configured, for example a traverse sub-system on a single disk could include the Initial Bearing, Bearings, Traverse, Lines and Trigonometry programs.

1.10.2   Hardware requirements

The minimum configuration for Version 2.1 is a microcomputer with screen and a single disk drive, although a dual drive is preferable. The screen may be 40 or 80 columns wide by 24 lines, all programs providing screen displays of data and results on a 40-column format. For hard copy printouts a standard dot matrix printer providing at least 80 characters per line on fan-fold pin-feed paper is required. The minimum page length is the standard 66 lines. Certain daisy-wheel printers taking cut sheets may be used, but this is not recommended due to the slowness of the typical daisy-wheel printer.

1.10.3   General advantages

The principal advantages in using a microcomputer for surveying calculations are speed and accuracy. When a program has been designed for a specific application, then it will be much faster in execution than an individual using a calculator. Further, if the program routines have been thoroughly tested then the accuracy of the calculations can be guaranteed (but not, of course, the accuracy of the surveying) since the machine will not make human errors of omission or forget the correct routines to be followed. For speed and reliability, of course, all software should be compiled in order to take full advantage of the machine's capabilities and avoid the risk of the user corrupting the programs.

Printer output of results is fast, standard formats may be used and such results are easily read and stored for reference.

The use of a floppy disk system allows observations and results to be stored in disk files, and these files may be rapidly recalled when errors in data have been detected and data entries must be amended. The saving in time in carrying out the computations may allow the early detection and rectification of mistakes in the fieldwork. Similarly, the possibility of amending values and re-running the computation in a few seconds will allow trial and error of different design parameters in, for example, the design of vertical curves.

The use of standard applications software may encourage the 'occasional surveyor' to use survey methods which he would not have considered worthwhile by hand computation methods, e.g. the bearing and distance methods now popular with electronic distance measuring equipment. It should be noted, however, that the use of an application program 'disciplines' the surveyor into executing the survey in a method suited to the particular program and there is a loss of individuality or freedom in designing the approach to the task.

1.10.4   Array sizes and operating systems

Commodore (CBM) operating system editions of MICROSURVEY include

(i) a 4000/8000 series edition, and
(ii) a CBM 64 edition.

Where the text quotes the number of lines, points, etc. which a programe can handle, these numbers refer to both CBM editions. These values are, of course, fixed by the array sizes set in the individual programs.
Standard operating system editions are available for

(i)   Z-80 based micros running under CP/M-80,
(ii) 16-bit micros running under CP/M-86, and
(iii) 16-bit micros running under MS-DOS or PC-DOS.

In these editions the array sizes are pre-set, but they may be re-configured by the user. As a general rule CP/M-80 issues are set to roughly half the array sizes used in CBM editions. For 16-bit machines the array sizes are set to at least twice the values used in the CBM editions.
Versions are also available for field use with hand-held microcomputers, the small micro being much cheaper than the data loggers supplied by many survey instrument companies.

Chapter 2

The Ordnance Survey

2.1 The national mapping agency

The *Ordnance Survey* ('the O.S.') has been the national mapping agency of Great Britain for 200 years. Other countries have similar agencies. The inexperienced surveyor should be aware of what the O.S. has to offer, not merely as regards the maps and mapping services they provide but also in respect of their collection and presentation of survey data.

It is sensible to emulate their work, since all survey work must be capable of being understood by others, and this is made easier if all surveyors adopt a common code of practice. The O.S. methods, however, are not the only forms of collecting and presenting data, and it would be undesirable to eliminate all personal characteristics in survey work.

2.2 Historical background

During the years 1745 and 1746 the armies of the Union were engaged in fighting the Jacobite supporters of Prince Charles. Troop movements, in the Scottish Highlands in particular, were considerably handicapped by the absence of any accurate maps of the country. The same problems arose in the 'pacification' of the Highlands, and as a result much of Scotland was mapped by the military, while at the same time a network of military roads was constructed. Following these mapping efforts, the military surveyors were employed on a scientific task, triangulation, designed to link the observatories at Greenwich and Paris. This led to the formation of the Ordnance Survey in 1791 as a military organization. An early task was the creation of a 1 inch to 1 mile scale map of southern England, a military (defence in this case) requirement during the Napoleonic wars.

After these wars, the need for maps was better appreciated, for example in the management and transfer of land, in civil engineering (railways, etc), in geological surveys, and in the survey of archeological sites.

Between 1840 and 1860 the Ordnance Survey began to assume its modern role of providing, as a national service, the surveys and maps required by Great Britain for military, scientific, commercial and industrial purposes. It provided maps at scales of 1 inch to 1 mile, 6 inches to 1 mile, 1:2500, 1:500 and, in the case of London, 1:1056. Following the 1914–18 war the staff was reduced to 1000 men (soldiers), and between 1925 and 1931 a number of Acts of Parliament became law, including Town Planning, Land Registration, Land Drainage, Local Government, Slum Clearance and Land Valuation Finance. The O.S., with its greatly reduced staff, could simply not cope with the work involved and as a result the Davidson Committee was set up in 1935 to review the situation.

It was already evident that more staff, a new triangulation and new surveys were needed. Recruitment of more civilians and the new triangulation commenced immediately, but the Committee's final report was not produced until 1938. The 1939–45 war then delayed the implementation of much of the Committee's recommendations, many of which still apply and are identified in the following sections.

2.3   The Transverse Mercator Projection

If the curved surface of the Earth is to be projected on to a flat sheet of paper some distortion must occur in one or more of the following: *shape* (orthomorphism), *direction* (bearings), *size* (areas as a quantity).

Clearly, then, scale cannot be consistent in all directions from all points on a map. Nevertheless, for many practical purposes the inconsistencies may be ignored, as is usual in plane surveying.

Prior to the Davidson Committee's recommendations, the O.S. maps were a series of county surveys, or combined counties, on a series of different projection data, the large-scale maps (1:2500, 1:1056 and 1:500) being known as the *County Series*. To the user who had property stretching across county boundaries it could be most disconcerting since adjoining counties did not always fit together. These discontinued plans and maps of the County Series may still be met in some offices.

The Committee recommended a single national projection for all new maps and this projection, a modified version of the *Transverse Mercator*, is used today. In order to appreciate a projection with reasonable simplicity, the reader should attempt to visualize a sheet of paper touching (or cutting) the surface of a globe at one or more points. The sheet may be flat, or wrapped around the globe to form a cylinder, a cone, part of a cone, or a spiral. If the lines of latitude and longitude are projected outwards (or inwards) from some point, say the centre of the Earth, on to the paper, then when the paper is laid out flat again some particular *map projection* will have been achieved.

Great Britain is a country whose length is greater in the north–south direction than in the east–west. One possible solution, therefore, is a cylindrical projection touching the globe along a meridian of *longitude* (a line on the surface of the Earth running between the North and South poles). This form is known as a *Transverse Mercator projection*. A Transverse Mercator projection is orthomorphic over any small area, scale is correct along the 'central meridian' (where the cylinder of paper touches the surface of the globe) and increases gradually to the east and west of the central meridian. To reduce the scale error at the east and west extremities of the country, the O.S. modified the projection by making the diameter of the cylinder a little less than that of the globe. This reduced the scale error at the extremities but introduced a scale error at the central meridian equal to, but of opposite sign to, that at the extremities. This results in there being two north–south lines (approximately 180 km east and west of the central meridian) where there is no scale error. The Earth actually approximates to an *oblate spheroid* (flattened at the poles) and in practice the lines of latitude and longitude are produced mathematically from formulae defining the projection and the shape of the Earth. (For further reading, see the Bibliography.) The *'origin of the projection'* (and of the calculations) is Latitude 49° North and Longitude 2° West (the *central meridian*).

2.4   The National Grid

The Davidson Committee recommended that a *National Grid* should be superimposed on all maps (subject to a few minor exceptions) to provide both a single reference system for the whole country and a means of numbering the proposed large-scale plans and medium-scale maps. The unit to be used for the grid was to be the *metre*.

Prior to the Committee's recommendation, War Office editions of O.S. maps had, for many years been overprinted with a metric grid to provide a means of identifying a point on a map. (The O.S. had also experimented with a grid based on the yard and 'kiloyard'.) Similar grids, usually in metres but occasionally in other units, are found on maps from all parts of the world.

The *National Grid (NG)* is an imaginary network of lines parallel to, and at right angles to, the central meridian of the projection, so forming a series of squares on the maps. The *origin* of the grid (the zero point) is southwest of the Scilly Isles and lies 400 km west and 100 km north of the origin of the projection, and it is therefore usually referred to as the *false origin*. This false origin means that every point in Great Britain has a reference consisting of a positive easting and a positive northing with respect to the grid origin. (It is a rectangular co-ordinate system in which the mathematician's x and y distances become 'Eastings' and 'Northings'.)

2.4.1   Recording the position of survey data

The position of instrument stations in the O.S. triangulation network or in one of their traverses are calculated and presented in NG units and hence other surveyors may plot or tie their own survey work to these O.S. positions. The values are quoted in metric units east and north of the false origin, e.g.

Manor Farm triangulation pillar
439 133.37 m East
120 988.83 m North

Note that Eastings are always quoted before Northings and the unit, m, is usually omitted. Tying traverses to O.S. control is referred to in Chapter 6.

2.4.2 A means of reference

The grid is sufficient for most people to devise a simple reference system for their own purposes and many users have done so. The preferred system, however, is as described below.

The NG is a systematic breakdown into squares of varying sizes. The largest unit of the grid is a square of 500 km side length, each square being identified by a letter as shown in *Figure 2.1*. Note that the letter 'I' is not used. Great Britain being a small country, most of its lies in the two

A	B	C	D	E
F	G	H	J	K
L	M	N	O	P
Q	R	S	T	U
V	W	X	Y	Z

Figure 2.1

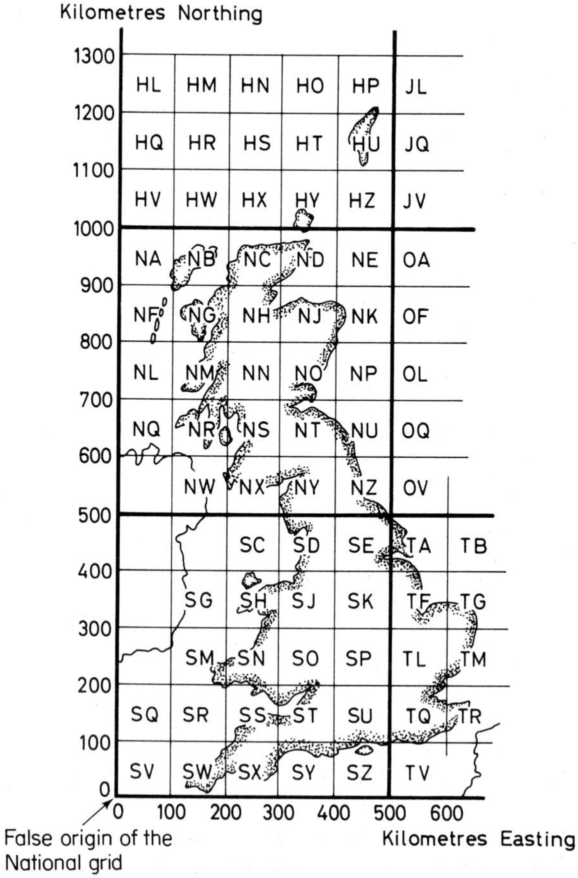

Kilometres Northing

Figure 2.2

False origin of the National grid

Kilometres Easting

squares 'N' and 'S', appropriately North and South, while a little, the Orkney and Shetland Islands, lies in square 'H' (Higher still). A little less, if any, lies in square 'O' (the Open sea), and the mouth of the Thames lies in square 'T'. The false origin of the NG is the southwest corner of square 'S'.

Each square of 500 km side is subdivided into 25 further squares, each with a side length of 100 km. The new squares within a 500 km square are similarly identified by letters as illustrated in *Figure 2.2* Thus, each 100 km side square is uniquely identified by a two-letter reference.

Any smaller square within a 100 km square is identified by the distances from the south-west corner of the 100 km square to the southwest corner of the smaller square, expressed in Eastings and Northings respectively. For example, SK 91 identifies a 10 km square and hence a 10 km reference, see *Figure 2.3*.

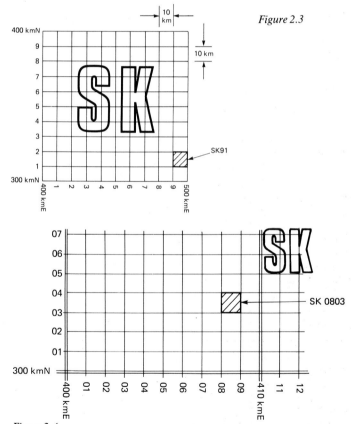

Figure 2.3

Figure 2.4

SK 0803 defines a 1 km square and hence a 1 km reference, see *Figure 2.4*. Similarly, SK 123456 defines a 100 m square, a 100 m reference, and SK 12345678 a 10 m square and a 10 m reference.

Note that every reference contains an even number of digits, half indicating the distance east and the latter half distance north. No digit should be omitted, a zero being used if needed as in *Figure 2.4*. If the two letters (SK in these examples) are included, the reference is unique and does not recur in the country. If the letters are omitted, the reference repeats in every 100 km square. If all work is confined to the same 100 km square and confusion cannot arise, the letters may be omitted. The grid lines appearing on maps depend upon their scale, see Section 2.5. When giving 10 or 100 m references, the final eastings and northings figures are often estimated.

2.4.3   Sheet numbering of the large-scale maps

The Ordnance Survey originally divided maps into large-scale plans, medium-scale maps and small-scale maps. In recent years this division has been amended to only two groups, the large-scale plans and medium-scale maps now being known, together, as large-scale maps, while the remainder are still described as small-scale maps. The new National Grid Series maps were to be square in shape if of a large scale, e.g. one square kilometre exactly, or 100 square kilometres (10 km by 10 km), having as sheet number the grid reference of the square, say SK 0803 or SK 91 respectively. The small-scale maps would remain oblong in shape, their numbering to be independent of the NG although the NG was to be superimposed on the maps. The relationship between the large-scale maps and their numbering is shown below.

2.5   Maps and map scales

The Davidson Committee recommended that the 1:2500, the 6 inch to 1 mile (1:10 560) and the 1 inch to 1 mile (1:63 360) scales be retained and that a new 1:25 000 and a 1:1250 scale should be considered. The 1:500 and 1:1056 scale maps referred to in Section 2.2 are no longer produced. In the mid 1970s, the 1 inch to 1 mile was replaced by a 1:50 000 scale and the 1:10 560 is slowly being replaced by a 1:10 000 scale. Both 1:1250 and 1:25 000 scale maps have been produced, although there was an attempt to cease production of the 1:25 000 some years ago.

The Ordnance Survey also produce maps at other scales, e.g. route planning and administrative maps at 1:625 000, motoring maps at 1:250 000, and archaeological and historical maps at scales from 1:2500 to 1:1 000 000. Some of the map series particularly relevant to the surveyor and others involved in survey tasks are referred to below.

2.5.1   1:50 000 scale (Landranger series)

This small-scale map is possibly the most popular of the O.S. scales, each sheet covering an area of 40 km by 40 km. It is not numbered with respect to the NG, but the grid is printed upon sheets at 1 km intervals, each 10 km line being shown in a broader gauge of line. This map is therefore useful as an index to all those large-scale maps using an NG reference as their sheet number. The 1:50 000 map is available in black and white (outline edition) as well as colour.

2.5.2 1:2500 scale

This series of large-scale maps covers the whole country except for mountain and moorland areas. It is also the scale of survey, except in the larger urban areas where the 1:2500 map is derived from the 1:1250 scale map. Originally each 1:2500 sheet covered an area of 1 square kilometre, its sheet number being the NG reference of the 1 km square, e.g. SZ 3178 (the grid values at the SW corner of the sheet). *Figure 2.5* shows a portion of a 1:50 000 sheet and how it may be used as an index for 1:2500 map.

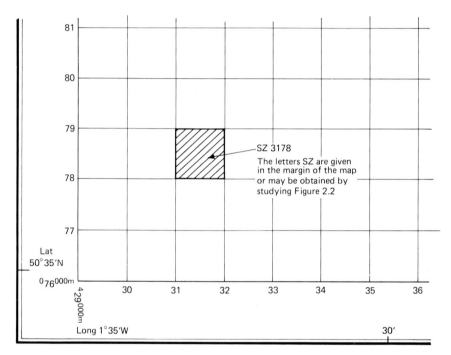

Figure 2.5

The County Series 1:2500 sheet covered an area of 1 mile by 1½ miles (approx. 1.6 by 2.4 km) and initially many users of the NG series were displeased by the small size of the new map. To overcome this objection the O.S. now regularly produces the 1:2500 scale map as a double sheet covering an area of two square kilometres. The double sheet is usually paired as in *Figure 2.6*, with the even numbered sheet being on the left, and the whole sheet would be known, for example, as SK 1234–1334.

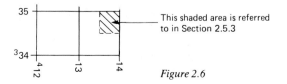

Figure 2.6

2.5.3   1:1250 scale

Twice the scale of the 1:2500 map, this is the largest scale map in regular production. It is the basic scale of survey in the larger urban areas. Each sheet is also a perfect square, using the same paper size as the original NG 1:2500 map. Hence the 1:1250 map covers one-quarter of the ground area of the 1:2500 map, that is to say 500 m by 500 m.

The 1:1250 scale map is identified by its 1 km reference and its quadrant within the 1 km square, i.e. NW, NE, SW or SE. For the shaded area in *Figure 2.6* the sheet number would be SK 1334 NE. This map is not produced as a double sheet. The NG on both 1:2500 and 1:1250 scale maps is at 100 m intervals.

2.5.4   1:25 000 scale

This series is the smallest scale of the large-scale maps and covers the majority of Great Britain. It was originally produced on a similar paper size to the 1:2500 scale map, hence each sheet covered an area of 10 km by 10 km, the sheet being known by its 10 km reference, e.g. SK 91. Like the 1:2500 scale, it is regularly produced as a double sheet, again usually paired horizontally with the even-numbered reference being on the left, e.g. SX 88/98. A number of special sheets at the same scale, called the Outdoor Leisure maps and covering an area of approximately 500 square kilometres, are also produced. These special sheets are named and not given a grid reference identification, although the NG is superimposed on them.

2.5.5   1:10 560 and 1:10 000 scale

This is another large-scale map series. It is identified by its grid reference in a similar manner to the 1:1250 scale map, since each sheet covers exactly one-quarter of a single 1:25 000 scale map (5 km × 5 km). An example reference is SX 88 NW.

The 1:10 560 or 1:10 000 scale maps cover the whole of Great Britain and these are the basic survey scales in some mountainous and moorland areas. The map is not produced as a double sheet, although special sheets of some towns and cities are available. The NG on both these scales and the 1:25 000 scale map is at 1 km intervals, hence these scales may also be used as an index to the 1:1250 and 1:2500 scale maps.

2.6   Continuous revision

A further recommendation of the Davidson Committee was that the large-scale maps should not be allowed to get out of date again and that a method of *continuous revision* should be adopted. When the 1:1250, 1:2500 re-survey areas and the 1:2500 county series maps were re-cast and revised on to the NG sheet lines, the O.S. started the practice of leaving one or more surveyors in the area, with the task of keeping the maps up to date. Most large towns have a small O.S. office which may be located from the telephone directory.

2.7   Ordnance Survey services

For many years the O.S. have provided a number of mapping services apart from the printing of maps on paper. It was announced in Parliament in 1973 that the O.S. would continue to produce and maintain the basic survey at 1:1250, 1:2500 and 1:10 000 scales. It would also make the survey information available in various forms appropriate to the needs of the users, but there would no longer be any general obligation to supply maps at 1:1250 and 1:2500 scales printed on paper. These maps, however, are still being produced on paper. The services described in the following sections are now provided.

2.7.1   SIM (Survey Information on Microfilm)

Today traditional paper copies of 1:1250 and 1:2500 scale maps are revised and reprinted only after considerable development of the survey area has taken place. To enable the map user to have more up to date information, the O.S. provides two alternatives.

The first of these is a 35 mm microfilm *copycard* from which a paper copy or a transparency may be produced at the scale of survey. Copies are also available at other scales, dependent upon the equipment used. This service is available from O.S. agents, there being one in most large towns. The agency is usually held by one of the good bookshops and/or stationers in the town. The agent's copy of the copycard is usually updated (replaced) some five times between the printing of traditional paper copies. The user may purchase his own microfilm copycard. Experience indicates that the quality of printout varies between the agents.

2.7.2   SUSI (Supply of Unpublished Survey Information)

SUSI is the second alternative, providing users with a more up to date service than SIM. The cost of the service is higher than that of SIM. Continuous revision is recorded on to Master Survey Drawings which are held at the local O.S. offices. Usually transparencies and paper copies of these drawings are obtainable on request from the local offices and enlargements can be obtained.

2.7.3   Digital mapping

Basic survey data are becoming available to users as *co-ordinates on magnetic tape*. Programs enable this information to be plotted in full, or part only, independent if necessary of NG map sheet lines and at a scale convenient to the user. It is possible also to merge the users own digitized information with that of the O.S. Developments are still in progress.

2.7.4   Reprographic services

The O.S. will supply enlargements and reductions of map sheets, part map sheets, or part combined sheets, usually as transparencies but also on paper. The completed work can be produced to the following scales –

1:500, 1:1250, 1:2500, 1:10000 or 1:10560. The image of the map may be printed as a forward or reverse reading. It should be appreciated that an enlarged map can be no more accurate than the map at the original scale. Transparencies are available of map sheets also at the scale of publication.

2.7.5   Aerial photographs

Some of the mapping of Great Britain has been carried out from aerial photographs. Contact prints, film positives, negatives and enlargements of this photography are available for purchase. the photography is black and white 'vertical air photography', see Chapter 15. The approximate scale of photography is usually either 1:7500 or 1:24000.

2.7.6   Control information

The horizontal and vertical (height) control required to map Great Britain includes *triangulation stations*, points established from traversing (known as *minor control points*) and height control (known as *benchmarks*). This control information is available for purchase. Horizontal control points information consist of a description, a photograph or sketch and the NG co-ordinates of points. In some cases also, details of ownership of land or the tenant of the land upon which the control point is located. Use of this horizontal control information is referred to in Chapter 6 on Traversing. The use of vertical control benchmarks, possibly the most widely used of the O.S. services, is covered in some detail in Chapter 8.

2.7.7   Other services

The O.S. is prepared to carry out survey work to individual specifications. This could be fieldwork, map mounting on cloth, calculations such as NG co-ordinates to Latitude and Longitude or vice-versa, etc. Approximately at monthly intervals the O.S. issues, on request and free of charge, a *Publication Report* listing maps and other survey documents published during the previous month.

2.8   Copyright

All users of O.S. maps, other services, or mapping based upon O.S. material, must obtain permission before producing copies in whole or in part. Crown copyright of O.S. publications remains in existence for 50 years. A copyright licence fee may be payable, royalties may be as little as 1p per copy. The reproduction of the NG on maps and other publications is permitted without reference to the O.S. and without payment provided an acknowledgement is given, for example 'The grid on this map is the National Grid taken from the Ordnance Survey map with the permission of the Controller of Her Majesty's Stationery Office'.

2.9   The Serpell Committee

The longer term policies, activities and financing of the O.S. were considered by this new committee in 1978. It reported towards the end of the year, having taken into account the views of map users and others as regards the requirements of a national mapping and survey service. No dramatic proposals were made, but there were recommendations to increase the area covered by the 1:1250 scale map series, and that map revision in some rural areas should be more up to date. Concern was also expressed in respect of marketing, revision techniques and man management.

Following the publication of the Serpell Report, the Secretary of State for the Environment of the day stated in Parliament that his aim was '. . . to see the Ordnance Survey flourish as an efficient and cost-effective organization, and to minimize its call on public funds while maintaining the quality of its work'. He also expressed the view that possibly the O.S. need not remain a Government department and that changes might be necessary in its financial structure. These statements caused some anxiety amongst map users, however, a Government move towards commercialization of the O.S. was defeated in the House of Lords in early 1983.

2.10   Interpretation of the O.S. 1:1250 and 1:2500 NG scale maps

As stated earlier, the surveyor should be aware of the methods of presentation of survey data used by the O.S., since this will guide him as to how best to present his own survey work and will also allow him to determine what information can be interpreted from the larger of the large-scale O.S. maps.

All the map sheets of both series are distinctive, yet they contain much uniformity, due to the use of a comprehensive set of principles and rules by both the field and the office staff of the O.S. when producing maps.

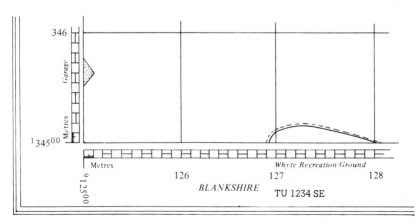

PLAN TU 1234 NE

Figure 2.7

2.10.1   Sheet size

1:1250 scale map – 500 m × 500 m.
1:2500 scale map – 1 km × 1 km or 2 km × 1 km.
A few sheets at both scales are extended in size at the edge of survey areas,
e.g. at the coast line.

2.10.2   The National Grid

The NG is shown on both map series at 100 m intervals, each grid square
representing one hectare (1 ha) in area, see *Figure 2.7*.

2.10.3   Margin and border information

Referring to *Figure 2.7*, the following are shown:

(1) National Grid values.
(2) A scale in tens of metres, completely surrounding the map detail.
(3) Sheet and adjoining sheet numbers.
(4) Names of features whose centre (area or linear feature) falls outside
the area of the map although part of the feature is contained within the
map.

The O.S. defines *names* to be – *administrative* (e.g. West Dorset County
Constituency); *distinctive*, being a proper name or noun other than
administrative (e.g. Gernick Estate); or *descriptive*, such as a common
noun (e.g. garage). Some or all of these names may appear within the
margins of a map.

2.10.4   Map detail

Detail is the name given in large-scale mapping to the natural and
man-made features on or adjacent to the surface of the Earth. One of the
principles referred to above is that the O.S. shall show all 'permanent
detail', in its true plan position as at ground level, although with a few
exceptions. This has led to the incorrect belief among a few map users that
these map series show all detail and that it is true to scale. However, some
detail is omitted, some is generalized and in some cases conventional signs
are used.
   A modern built-up area may be extremely complex – consider any city
centre area the reader may be familiar with and the problems of any
surveyor or draughtsman attempting to map or draw such a place will be
self-evident. For example, what if the detail is too small to be drawn to
scale, or if a jut, recess or porch on a building is too small to be plotted to
scale? Again, when two parallel features are so close that the draughtsman
cannot show both of them with clarity and in their true plan position. In
such cases, the surveyor has the choice of not showing the features, using
conventional signs, or generalizing the detail.
   Again, what is meant by 'ground level'? In a pedestrianized underpass
with shops and a sunken garden under an existing network of roads, the
road network might be considered to be at 'ground level'. However, if the
pedestrianized area was at an original ground level and the roads were

Figure 2.8

elevated to pass over the shops, then the shops might be considered to be at 'ground level'. Yet again, the shops may have been partially sunk and the roads partly elevated, with respect to an original ground level.

The O.S. considers all three cases to be similar and overcomes the problem with the maxim '. . . if the ground level is not self-evident then it shall be defined as "the upper level of surface communication" . . .'. However, some underground detail is shown, for example in the above situation the outline of the pedestrianized lines of communication (footway or pavement) would be defined. Similarly, some overhead detail is shown, particularly if it is considered to 'constitute a useful feature of the map'. Such examples include cantilevered bridges and balconies, buildings supported on pillars or columns, etc.

Detail is represented by three classes of conventional signs – *line* symbols, *point* symbols and *area* symbols. All linear detail at ground level which might constitute an obstruction to an able bodied pedestrian (walls, fences, hedges, water's edge, railway lines, etc.) are shown by firm lines, while other linear detail (overhead and underground detail, edge of footpaths, grass verges, etc.) is shown by pecked lines. A *pecked line* is a form of broken or dashed line (see *Figure 2.8*).

Point symbols are used to represent detail which is too small to show to scale (see *Figure 2.8*).

Area symbols are used to define roofed-over areas, a glazed roof by cross-hatching and other roofs by stippling (dots). Areas of vegetation and surface features depicted by area symbols are shown in *Figure 2.8*.

Names (as described above) whose centre falls within the body of the map are shown, together with postal (house) numbers of buildings, if any. Where numbers follow a regular sequence, sufficient only are shown in order to enable the user to identify the remainder. The numbers are printed with the base of the number facing the street to which they refer, unless lack of space dictates otherwise. Horizontal and vertical control points are also shown, e.g. minor control points either as .rp (*revision point*) or .ts (*traverse station*). *Triangulation stations* are shown as a dot within a triangle, and *benchmarks* as a 'broad arrow' followed by, for example, BM 56.64 m. Surface heights of some roads are indicated, to 0.1 m, e.g. + 27.2 m (see *Figure 2.8*).

A further principle is that all administrative boundaries be shown, these being depicted by a range of line symbols and annotation. A descriptive abbreviation is given of the area defined, e.g. 'Co Const Bdy', that is to say a Parliamentary County Constituency Boundary. In addition, an indication is given of the detail to which the boundary is related. A boundary is said to be *mered* and to relate a boundary to a length of detail is called *mereing*. The study of a map indicating administrative boundaries will show many mereing abbreviations, most of which are relatively self-explanatory. For example, Und, Def, CC, CR, FW, TB, 1.22 mRH mean, respectively, undefined (boundary not related to any detail visible), defaced (detail to which the boundary was originally mered no longer exists), centre of canal, centre of road, face of wall, top of bank and 1.22 m (4 ft) from the root of a hedge.

Mereing symbols are used to define the position of a change of mereing, see *Figure 2.9*.

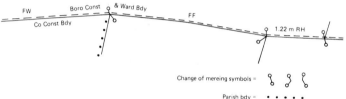

Figure 2.9

2.10.5   Parcels and areas on the 1:2500 scale map

Apart from the difference in scale, 1:2500 maps differ from the 1:1250 in that in rural localities the area of each 'parcel' of land is shown. A *parcel* of land is defined as a distinct portion of land, although in some cases the O.S. have grouped small portions of land together into one parcel. Each parcel on a map bears a reference number, together with its area in hectares and/or acres.

For example

1793
1.763 ha
4.29

Here, 1793 is the parcel reference number, generally a simple 10 m reference to its approximate centre. Exceptions to this rule arise where parcels lie across map edges. Where a parcel lies across a corner of maps, it is allotted a number 0001 to 0006 such that no number is repeated in any one kilometre sheet and the number used is common to the four map sheets upon which the parcel lies. A parcel which lies across map edges but does not cross the corners is given the 10 m reference of the centre of that portion of the map edge passing through the parcel. Where parcels lie across map edges, their areas are quoted to the map edge only.

A parcel was defined above as a distinct portion of land, but this may not always appear so when studying an O.S. 1:2500 map. A parcel may consist of a group of small portions of land, or on occasion, be subdivided by another parcel and sometimes the boundary of a parcel may be the centre of a double feature. To assist in resolving these situations, the O.S. use a map symbol known as a *brace*, similar to a long S or the integration symbol. Brace symbols may be of four types – single brace, double brace, open brace, or centre brace.

Single and *double braces* are used to symbolize the linking together of two or more portions of land to form a single parcel. An *open brace* is used to tie together the parts of a parcel which itself is divided by another parcel, e.g. a bridge across a river, or an island in an ornamental lake where the island is considered to be part of the adjoining land area. A *centre brace* is used to define the edge of a parcel which is the centre of a 'double' feature such as adjacent and parallel fences or hedges, see *Figure 2.10*.

Figure 2.10 also shows further map symbols which look like a line of map pins. This line symbol defines the edge of a build-up area which is assumed by the O.S. to be a single parcel. No braces appear within this area.

32

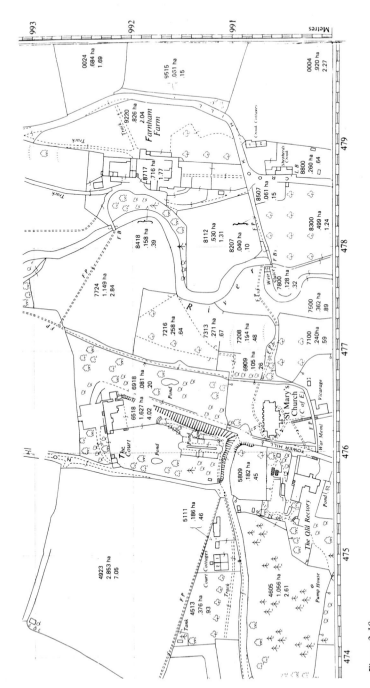

Figure 2.10

Part 2

Traditional site surveys

Chapter 3

Site investigations

3.1 Purpose

Site investigations are concerned with the assessment of the suitability of an area of land for the construction of building or civil engineering works. These investigations involve the acquisition of knowledge of the factors which will affect the design and construction of such works and the security of adjacent land and properties. The findings of the site investigation will influence decision making as to whether the proposed development of the site is feasible in technical and economic terms.

3.2 Primary objectives

The primary objectives of a site investigation are set out in BS 5930: 1981 British Standard Code of Practice for Site Investigations. These may be summarized as follows:

(1) To assess the suitability of the site for the proposed works.
(2) To enable an adequate and economic design to be prepared.
(3) To identify factors which may create difficulties during the construction process, in order that they may be planned for and the best method of construction chosen.
(4) To determine the effect of changes, either natural or arising from the proposed works, on adjacent buildings, the local environment and the proposed works themselves.
(5) Where alternatives exist, to evaluate the relative suitability of sites or parts of the same site.

3.3 Initial site appraisal

Much information concerning a site under consideration may be available from existing records and enquiries should be made to determine whether the owner of the site or his agent has any such information.

Appendix B of BS 5930 lists several sources of published information which may prove helpful. These include O.S. maps, Institute of Geological Sciences maps and memoirs, Meteorological Office reports and statistics and the Department of the Environment Central Register of Air Photography of England and Wales.

The local authority responsible for the area within which the site is situated can often provide useful information, and copies of old maps, etc., are frequently held in the archives of local libraries, county record offices and museums and are available for public inspection.

The statutory undertakers responsible for providing utilities such as gas, water and electricity should be consulted as to the location and capacity of their mains to service the site. It is helpful if a letter enquiring into these matters is accompanied by a plan of the site and its immediate environs (perhaps a photocopied extract from the relevant O.S. sheet) for them to 'mark up' with the relevant information and return.

Every effort should be made by the client's legal advisers to establish any matters such as easements (the rights which one owner of land may acquire over the land of another) or restrictive covenants (an obligation that prevents the person who takes the land with notice of the covenant and his successors in title from doing something on their land, e.g. building more than one house) which might frustrate the proposed development, at the earliest opportunity to avoid further, perhaps abortive investigations. However much information can be collected off-site, an early reconnaissance is advised. Appendix C of BS 5930 contains a series of useful notes on site reconnaissance.

During the initial inspection it is invaluable to have a camera handy in order to prepare a photographic record which can then be studied later, together with other data collected in the preparation of a feasibility report. The time available on site is often limited and the camera offers a very quick and accurate method of recording information which will be readily assimilated by others who have not visited the site. A camera with a 'data back', i.e. a camera back containing a quartz clock which will automatically record time and date by exposing these by means of a tiny internal flash at the bottom of the frame each time a picture is taken, is a most useful device.

3.4   Initial site appraisal check lists

The following sections provide check lists for the initial site appraisal. After this has been carried out the extent of the additional detailed information required will be clearer. This additional information may include a detailed site survey, surveys of existing buildings on site and sub-surface investigations.

3.4.1   Site access check list

Classification of roads serving the site, together with their general condition.

Road widths, traffic flow and restrictions, e.g. low bridges, weight limits.

Positions of road intersections and discharge points (e.g. factory gates) which may be potential hazards.

Existing vehicular and pedestrian access points to the site and visibility at junctions.

Footpaths and rights of way.

3.4.2   Services check list

Drainage and sewerage: location and level of existing systems (identifying whether foul, stormwater or combined), pipe sizes, existing flow and ability to take additional flow. Check signs of liability to surcharging or flooding.

Water supply: location, size and depth of main, pressure available.

Electricity supply: location, size and depth of mains, voltage, phases and frequency. Capacity to supply additional requirements. Transformer requirements.

Gas supply: location, size and depth of main. Pressure. Capacity to supply additional requirements.

Telecommunications: location of existing lines and capacity to supply additional lines.

Heating: availability of fuel supplies, district heating, smokeless zones.

Positions of poles, pylons and overhead lines (state safe clearance below).

3.4.3   Topography check list

Site boundaries and how defined, e.g. fence, hedge, ditch.

Positions, types and sizes of trees and hedgerows. Check with Local Planning Authority for Tree Preservation Orders.

Types and differences in vegetation (may indicate changes in soil conditions).

Ground contours and general drainage features.

Mounds or hillocks (natural or man-made?).

Positions of ponds, watercourses and wet patches. Liability to flooding.

Presence of cuttings and excavations or other signs of workings subsequently filled.

3.4.4   Underground hazards check list

Check with the local authority and other statutory bodies for mine workings, tunnels, water table, springs and ground movement, underground services.

3.4.5   Environmental check list

Orientation.

Degree of exposure and local climatic or other hazards by sea air, pollution.

Meteorological information: direction and strength of prevailing winds, annual rainfall and seasonal distribution, temperature range (seasonal and daily), severity and incidence of storms, liability to fogs, etc.

Presence of undesirable features, e.g. noisy or smelly adjacent land uses.
Adequacy of local facilities, e.g. transport, shops, schools, etc.

Photographs: include general views, views from known points, panoramic views, views of specific features (e.g. adjacent buildings, buildings on site, dominant features, particularly good or bad outlooks). Viewpoints should be recorded on a rough sketch or O.S. map extract and a note made of the date and time of day taken.

3.4.6   Legal and statutory check list

Names and addresses of: vendors; their agents and solicitors, neighbouring landowners; local authority; statutory undertakers for the area (e.g. gas, water, electricity).

The client's actual or potential legal interest in the land – freehold or leasehold.

Ownership of boundary fences and walls.

Easements: such as rights of way, rights of support to neighbouring land and buildings, right to light and air.

Restrictive covenants.

Existence of any valid Town and Country Planning Consents for the site.

Local Planning Authority policy for the area.

Features giving rise to special planning restrictions, e.g. Ancient Monuments, buildings of architectural or historic interest, burial grounds.

3.5   Detailed site survey

The measured survey of the site should be plotted to an appropriate scale and should possibly show the following physical details:

Precise boundaries.
Rights of way across the site.
Position of gates and access roads.
Hedgerows.
Position of trees, noting girth and spread.
Ditches and watercourses.
Ponds and wet areas.
Rock outcrops.
Hillocks.
Ancient monuments.
Bench marks, levels and contours.
Drain runs, levels and invert levels.
Electricity and telegraph poles, positions of overhead lines including a note of clearance or headroom available.

3.6   Surveys of existing buildings – condition surveys

In addition to preparing the detailed measured drawings as described in Chapter 7, an inspection and report on the condition of the building may also be required. To carry out such an inspection requires not only a sound understanding of construction technology but also a good knowledge of common building defects and their causes. A detailed consideration of these subjects is beyond the scope of this text, but the following sections may provide general guidance.

3.6.1 Scope of the survey

Although the survey should concentrate on the main structure, i.e. roof, walls and floors, some consideration should also be given to the services and the site. Every effort should be made to inspect as much of the building as possible.

Care should be taken to think of the property as a whole, and to consider individual defects as symptoms which may be inter-related and indicative of less obvious but more serious problems. The connection (and the differences) between cause and effect must always be borne in mind.

3.6.2 Equipment

The following basic equipment is recommended:

A4 Millboard and paper, or prepared check list forms.
Moisture meter.
Handlamp (plus spare bulb and battery).
Probing instruments (e.g. cheap screwdrivers).
Section ladder (3 m length).
Protective clothing.
Measuring and other equipment as in Chapter 7.

3.6.3 Inspection check lists

The next sub-sections provide check lists for general guidance as to those points to which the surveyor should pay particular attention.

3.6.3.1 Roofs externally

(i) Pitched roofs
Slopes, ridges and hips for signs of deflection or movement which might indicate failure of the structural framework.
Coverings for missing or defective slates/tiles which could permit water penetration, and for regularity of the course (irregular courses may be due to failure of the battens or nails).
Ridge and hip tiles, noting the condition of mortar pointing and bedding.
Leadwork (or other metals) particularly to valley and parapet gutters.
Eaves and gable details for defects, e.g. rotting fascia, soffit and barge boards, defective verge pointing.
(ii) Flat roofs
Adequacy of falls to dispose of rainwater (ponding may indicate deflection of the roof structure).

Upstands and abutments for cracks and the adequacy of the detailing (e.g. sufficiently high, properly flashed).

Coverings for splits, bubbles, cracks and crumbling asphalt.

(iii) Chimneys

Safety and stability – plumbness, flaunching to pots, etc.

Detailing at the junction with the roof covering, e.g. flashings, aprons, back gutters, etc.

Condition of pointing/rendering.

(iv) Rainwater disposal

Adequacy of design (e.g. size and position of gutters, r.w.p.s. and outlets, correct falls).

Gutters and outlets for freedom from obstruction. Joints for leakage (stains and dampness on adjacent walls).

Rainwater goods for cracking and corrosion.

3.6.3.2   *Roofs internally*

(i) Roof void (where accessible)

Adequacy of the structural frame (sizes, arrangement and connection of members as related to span/loading; bearings of purlins and rafters).

Coverings, battens and nailing (where not obscured by underslating felt).

Signs of damp penetration, particularly at chimney stacks, soil and vent pipes, abutments and beneath valleys.

Timbers for infestation by wood-boring beetles or wood-rotting fungi.

Thermal insulation and lagging to cisterns and pipework.

Signs of condensation, inadequate insulation.

(ii) Ceiling below

Signs of dampness and staining.

3.6.3.3   *Main walls externally*

Adequacy of foundations: although foundations are not normally exposed, an opinion as to their adequacy can be based on indirect evidence by examination of the visible superstructure, e.g. leaning, bowing and cracking often indicate foundation problems (where walls lean outwards be sure to check the bearings of structural members resting on the wall).

Walls for signs of structural movement, e.g. cracks, walls out of plumb, distorted window and door openings.

Wall surfaces for defective pointing, porous and soft brickwork, cracked or otherwise defective rendering which may allow water penetration.

Bonding of walls at abutments, particularly extensions to older work.

D.p.c. (if present) for continuity, and the relative levels of the d.p.c. to the surrounding ground and ground floor levels.

Air bricks ventilating suspended timber ground floors for adequacy and freedom from obstruction.

Copings and pointing to parapet walls.

Windows: check for rot in timber windows, particularly cills, and check for corrosion in metal windows.

External decoration.

3.6.3.4 Main walls internally

Wall surfaces for signs of dampness, check effectiveness of d.p.c. with a moisture meter, check that detailing around openings in external walls does not permit water penetration.

Walls and openings for signs of structural movement, e.g. cracking in plasterwork (not to confuse shrinkage cracks with cracks of a structural significance), split or stretched wallpapers, distorted door and window openings.

Windows to see that they are operational, open windows at upper levels and lean out to inspect the condition of the window externally and the condition of surrounding brickwork, rainwater goods, roofs at lower levels, etc.

3.6.3.5 Floors

(i) Timber floors

Signs of collapse, sponging, sagging or unevenness, with particular vigilance when inspecting suspended ground floors in older properties without effective d.p.c.s or adequate under-floor ventilation.

Exposed surfaces for signs of timber decay and beetle infestation.

(ii) Solid floors

Surface for signs of dampness evidenced by, for example, presence of salts, detached tiles, moisture visible when covering rolled back.

Surface for signs of movement, e.g. settlement on poorly compacted hardcore (fill deeper than 600 mm is prone to this form of failure), lifting from sulphate attack (note brickwork at the perimeter walls may be displaced laterally at d.p.c. level).

(iii) Ceilings

Surface for signs of cracking, deflection, detachment of lath and plaster.

3.6.3.6 Joinery

Exposed surfaces for evidence of wood-boring beetle, wood-rotting fungi and excessive warping and shrinkage.

3.6.3.7 Services

A cursory examination should be made, looking for signs of inadequacy.

Old electrical wiring (25 years plus).

Leakage from sanitary plumbing, heating and hot and cold water services.

Blockage or leakage in the drainage system: lift manhole covers, look for soft ground and depressions adjacent to the drain runs.

3.7   Schedule of condition

The information gathered from the inspection of the building may be presented in a report for the client's consideration, but it is often more convenient to present it in the form of a schedule as in the following extract:

Schedule of condition
'Akin Arms', Glass Street, Drinkfield
11 th August 1984

Elevation	Item	Description	Cond'n	Defects
Front(N)	Roof	Pitched, interlocking conc. tiles with half round ridge.	3	Mortar joints between ridge tiles failed. 6 No. tiles cracked.
	Ch'y	Brick, terra cotta pot.	4	Apron flashing loose and torn. Mortar joints eroded.
	R/W goods	100 mm half round pvc gutters, 65 mm pvc down pipes.	1	Nil, recently renewed.
	Wall	270 cavity, facing brick to ground storey and conc. vertical tile hanging at 1 st floor level.	3	Slight settlement cracking below LH gd. floor window (inactive), 20 No. cracked-slipped tiles.
	D.p.c.	Bitumen felt.	3	Discontinuity to RH of saloon bar entrance, with damp staining extending 150 mm above d.p.c. level.

Note: As a time-saving device the condition of the various elements and finishes may be described by reference to a numbered scale, e.g.

1 Very good
2 Good
3 Fair
4 Poor
5 Very poor.

If such a scale is used it must be defined in the schedule.

3.8   Sub-surface investigations

The initial site appraisal should have provided sufficient information to highlight any ground problems which will need to be investigated before the design and construction can proceed. These problems should then be the central considerations of a sub-surface investigation.

3.8.1   Soil investigation methods

For most low-rise small-span structures, e.g. housing, trial pits provide the best method of investigating the subsoil since they allow a very detailed examination of near-surface deposits. Trial pits are cheap, usually being dug using a hydraulic backactor machine such as a JCB or a Hymac.

The pits are normally approximately 1.5 m² in plan and excavated to a depth at least one and a half times the width of the proposed base below the level at which the base is to be founded. For example: a typical house wall strip foundation of 600 mm width, which is to have its base set 900 mm below finished ground level, would require a trial pit of 900 + (1.5 × 600) = 1800 mm depth for adequate information to be gathered to design the foundation. Pits should be adequately supported to prevent collapse by means of temporary timber or metal planking and strutting and guarded by a clearly visible barrier at ground level. The maximum depth to which trial pits are dug is generally in the region of 4 m, beyond this depth they become impracticable and expensive.

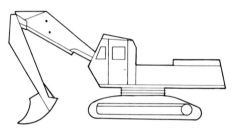

Figure 3.1

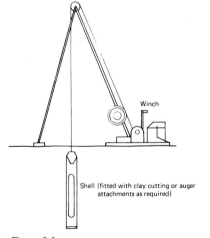

Shell (fitted with clay cutting or auger
attachments as required)

Figure 3.2

Where medium to high-rise developments are proposed, investigations at greater depths than 4 m are usually necessary. This may be done by forming boreholes. These are commonly made using a lightweight, towable shell and auger rig which will mechanically produce holes of 150 to 200 mm diameter to a depth of 45 m in clay. In sandy soils and soft strata the borehole may collapse if the wall of the hole is not supported. The most commonly used method to prevent this is lining the hole with steel tubes or casings. A borehole in sandy soils using 300 mm tools and casings can achieve depths of up to 60 m.

Soil investigations and the interpretation of the findings is a matter best left to geotechnic specialists, therefore the following sections are intended only as a general outline.

3.8.2   Soil profiles

From the trial pits or boreholes a soil profile can be established and a drawing produced showing the depth and composition of the various soil layers.

In describing soils a standard method of classification is used, as follows:

Soil type: sand, clay, organic, clayey sand, silty sand, etc.
Soil compactness: loose, firm, fissured, etc.
Soil colour, brownish-grey, reddish, black, etc.
Moisture conditions: wet, damp, dry, any seepage.
Roots: describe size, types and at what depths.
Fill: note content and mixture, type of any rubbish present, smell and colour.

3.8.3   Soil identification

General advice on soil identification is contained in Building Research Establishment Digest 64, *Soils and Foundations Part 2*, Table 1, Soil Identification. The table is reproduced below.

3.8.4   Bearing capacity

The foundation of a building serves to distribute the loads to be carried over a sufficient area of bearing surface so that the subsoil will be

TABLE 3.1. Soil identification

Soil type	Field identification	Field assessment of structure and strength	Possible foundation difficulties
Gravels	Retained on No. 7 BS sieve and up to 76.2 mm	Loose – easily removed by shovel	Loss of fine particles in water-bearing ground
	Some dry strength indicates presence of clay	50 mm stakes can be driven well in	
Sands	Pass No. 7 and retained on No. 200 BS sieve	Compact – requires pick for excavation. Stakes will penetrate only a little way	Frost heave, especially on fine sands
	Clean sands break down completely when dry. Individual particles visible to the naked eye and gritty to fingers		Excavation below water table causes runs and local collapse, especially in fine sands

Silts	Pass No. 200 BS sieve. Particles not normally distinguishable with naked eye	Soft – easily moulded with the fingers	As for fine sands
		Firm – can be moulded with strong finger pressure	
	Slightly gritty; moist lumps can be moulded with the fingers but not rolled into threads		
	Shaking a small moist lump in the hand brings water to the surface		
	Silts dry rapidly; fairly easily powdered		
Clays	Smooth, plastic to the touch. Sticky when moist. Hold together when dry. Wet lumps immersed in water soften without disintegrating	Very soft – exudes between fingers when squeezed	Shrinkage and swelling caused by vegetation
		Soft – easily moulded with the fingers	Long-term settlement by consolidation
	Soft clays either uniform or show horizontal laminations	Firm – can be moulded with strong finger pressure	Sulphate-bearing clays may attack concrete and corrode pipes
		Stiff – cannot be moulded with fingers	
	Harder clays frequently fissured, the fissures opening slightly when the overburden is removed or a vertical surface is revealed by a trial pit	Hard – brittle or tough	Poor drainage
			Movement down slopes; most soft clays lose strength when disturbed
Peat	Fibrous, black or brown	Soft – very compressible and spongy	Very low bearing capacity; large settlement caused by high compressibility
	Often smelly	Firm – compact	
	Very compressible and water retentive		Shrinkage and swelling – foundations should be on firm strata below
Chalk	White – readily identified	Plastic – shattered, damp and slightly compressible or crumbly	Frost heave
			Floor slabs on chalk fill particularly vulnerable during construction in cold weather
		Solid – needing a pick for removal	
			Swallow holes
Fill	Miscellaneous material, e.g. rubble, mineral, waste, decaying wood		To be avoided unless carefully compacted in thin layers and well consolidated
			May ignite or contain injurious chemicals

TABLE 3.2.

Minimum width of strip foundations

Type of subsoil	Condition of subsoil	Field test applicable	Minimum width in millimetres for total load in kilonewtons per lineal metre of loadbearing walling of not more than—					
			20 kN/m	30 kN/m	40 kH/m	50 kN/m	60 kN/m	70 kN/m
(1)	(2)	(3)	(4)	(5)	(6)	(7)	(8)	(9)
I Rock	Not inferior to sandstone, limestone or firm chalk	Requires at least a pneumatic or other mechanically operated pick for excavation	In each case equal to the width of wall					
II Gravel Sand	Compact Compact	Requires pick for excavation. Wooden peg 50 mm square in cross-section hard to drive beyond 150 mm	250	300	400	500	600	650
III Clay Sandy clay	Stiff Stiff	Cannot be moulded with the fingers and requires a pick or pneumatic or other mechanically operated spade for its removal	250	300	400	500	600	650
IV Clay Sandy clay	Firm Firm	Can be moulded by substantial pressure with the fingers and can be excavated with graft or spade	300	350	450	600	750	850

Type / Soil	Consistency	Description		
V Sand Silty sand Clayey sand	Loose Loose Loose	Can be excavated with a spade. Wooden peg 50 mm square in cross-section can be easily driven	400	600
VI Silt Clay Sandy clay Silty clay	Soft Soft Soft Soft	Fairly easily moulded in the fingers and readily excavated	450	650
VII Silt Clay Sandy clay Silty clay	Very soft Very soft Very soft Very soft	Natural sample in winter conditions exudes between fingers when squeezed in fist	600	850

Note: In relation to types V, VI and VII, foundations do not fall within the provisions of regulation D7 if the total load exceeds 30 kN/m prevented from spreading, and also to avoid excessive or unequal settlement of the structure. To enable the foundation to be designed the bearing capacity of the subsoil must be determined. Methods available include:

Simple field assessment. Reference may be made to the Table to *Building Regulation D7* which tabulates subsoil types, field tests, and suitable minimum foundation widths for residential buildings. Note that the average loading for a two-storey dwelling of traditional construction is of the order of 30–50 kN/m run (3–5 tonnes/m).

Site testing using specialist equipment such as the dynamic penetration test, or vane testing.

Taking samples for analysis and testing off-site in a soils laboratory.

BS 5930 Table 4 provides a schedule of laboratory tests on soil which are described in detail in BS 1377:1975: *Methods of Testing Soil for Civil Engineering Purposes.*

3.8.5   Shrinkable soils

Clay soils are prone to seasonal moisture content changes leading to shrinkage (in summer) and swelling (in winter). In order to avoid such dimensional changes disrupting foundations, they must be placed at a depth below the zone of seasonal movement. A measure of the shrinkability of clay is its plasticity index, which may be determined by laboratory test. Problems arise where clay soils have a plasticity index in excess of 40 and an index of 50 is considered a high risk necessitating deep foundations.

3.8.6   Harmful groundwater

If a soil analysis is being carried out it is normal also to analyse groundwater. Concrete foundations may be attacked by acidic or sulphate-bearing waters. The relative acidity of groundwater is indicated by the pH value; the lower the pH then the higher the acidity (the pH value for distilled or neutral water is 7). Sulphate content may be expressed in grammes of sulphur trioxide (SO_3) per litre in a 2:1 water:soil extract. Where there is more than 1 gramme per litre, then sulphate-resistant concrete mixes must be employed (see BRE Digest 250). Concrete made from ordinary Portland cement has no resistance to acid attack and concrete made from sulphate-resistant cement has only slight resistance. If acidic conditions are encountered then BRE Current Paper 23/77 *Chemical Resistance of Concrete* should be consulted for guidance on the design of chemically-resistant concretes.

Chapter 4

Chain survey

4.1 Introduction

Chain survey was mentioned earlier as a method of supplying *detail*, with *offsets* being taken from a framework of measured straight lines known as *chain lines*. The framework can be controlled by triangulation, or alternatively by traversing, or it may simply be a network of measured straight lines which are plotted to scale. This chapter is primarily concerned with the last technique, tying to traversing is referred to in Chapter 6. Offsets are also often used in setting out – see Chapter 17.

This chapter describes the equipment to be used and the field and office procedures of chain surveying with respect to the supply of detail for the production of plans at scales of 1:500 or less (e.g. 1:5000). The method is not recommended at scales greater than 1:500 because the accuracies usually attained in the alignment (*ranging*) and measurement of lines are generally inadequate at the larger scales (e.g. 1:200).

If accuracy is a criterion, and it often is, then the following limits of measurement are suggested:

Scale	Chainage and measurement of detail to nearest
1:2500	0.25 m
1:1250	0.1
1:1000	0.1
1:500	0.05

Note the similarity of the numerals in the two columns – it makes it easier to remember these figures.

The limits of measurement suggestions are based on the fact that it is possible to attempt to plot to 0.1 mm on the drawing material, therefore it should be correct within 0.2 mm. Ground measurements, therefore, need not be better than one can attempt to plot – for example, 0.1 mm on the paper at 1:2500 scale represents 2500×0.0001 m on the ground, i.e. 0.25 m, and similar values can be deduced at other scales.

The name *chain survey* results from the fact that the chain was the standard measuring device used in linear surveys for over two centuries. Before its invention in the seventeenth century by Aaron Rathbone, lines were measured with ropes, cords or wooden rods, in the same way as used by the Egyptians thousands of years ago.

4.2   Theory of chain survey

If the three sides of a triangle are marked out on the ground and measured, then the triangle may be plotted to scale on paper using a scale-rule and a pair of compasses. One line can be drawn to scale, then from its ends arcs may be drawn to the scale lengths of the other two sides. If all the measurements are correct, the intersection of the arcs defines the third 'corner' of the triangle.

In chain survey, a number of straight lines are laid out in the area to be surveyed, arranged so that they form a framework of triangles. If all the triangle sides are measured, then the framework can be reproduced on a plan.

Detail in the area is located by measurements from the lines of the framework. These, in turn, are plotted to scale on the plan until the plan gives a proportional representation of the area surveyed. The detail is 'picked up' by offsets from the chain lines, these lines sometimes being termed *detail lines*.

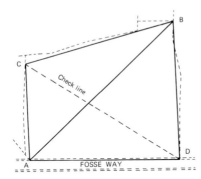

Figure 4.1

The triangle technique is simple and direct, but it has one important defect. This is that should an error be made in measuring the field length of one side of a triangle, then the triangle can probably still be plotted, but it will not be the same triangle as exists upon the ground. Errors in measurement are easily made – misreading figures, chain not straight, etc. – therefore it is essential that an extra line be measured in every triangle to act as a check on the measurement of the sides. The *check line* is not used in plotting the triangle, but after plotting – its length is then scaled off the drawing and compared with the measured field length. If the two agree it proves that there is no serious error of measurement or plotting of the triangle. *Figure 4.1* shows an example of chain lines which could be used to survey a field.

When a survey consists of several triangles, they must not be simply 'tacked-on' to one another, since this leads to a gradual build-up and exaggeration of any errors. Instead, one long line (as long as possible) should be placed right through the area of the survey and all the triangles based on this base line. The *base line* ties the whole framework together and prevents any gradual twist or deformation of the framework. It is, of course, the chain survey method of satisfying the rule, *work from the whole to the part*. The 'art' of chain survey lies in proper selection of the framework lines, see Section 4.5.1 on reconnaissance and the selection of lines.

4.3   Equipment

The following sections describe the equipment used for 'chaining' survey lines, and for the measurement of detail, together with the common small hand instruments. All of these are used in other branches of survey work besides chain survey. In addition, suitable clothing is necessary, and on occasion equipment for clearing vegetation and making concrete markers. The smaller items listed may be conveniently carried in a haversack.

4.3.1   The chain

Several forms of chain are available, according to the unit of measurement in use. The form, illustrated in *Figure 4.2*, is the 20 m chain specified in BS 4484:1969, which is the type recommended for carrying out 'chaining'.

Figure 4.2

The *chain* comprises 100 steel wire links, each of which is joined to its neighbours by two or three oval rings. Swivelling brass handles are fitted to each end of the chain and its total length of 20 m is measured from the outside of one handle to the outside of the other. *Figure 4.3* shows a part of such a chain.

To enable the chain to be read with ease, yellow plastic 'tallies' (*Figure 4.4*) are fitted at every metre except at 5, 10 and 15 m where red numbered tallies are used (*Figure 4.5*). Note that 5 m tallies are used at both the 5 m and 15 m positions, thus when laid upon the ground the chain may be used to measure in either direction.

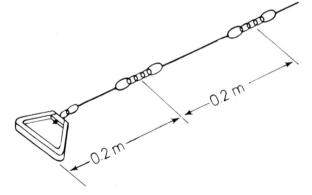

Figure 4.3

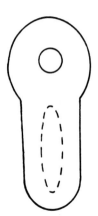

Figure 4.4

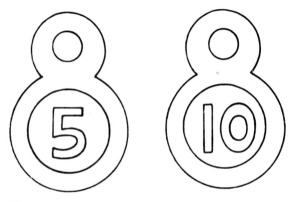

Figure 4.5

An individual reading of a 'chainage' figure may be read by estimation to 0.05 m. A line measured with a chain is, at best, possibly accurate to 1:1000, that is to say correct to one metre in a thousand metres measured, or to 0.1 m in every 100 m measured. In practice, however, chain measurements are often correct only to 1:500 or less, thus it is obvious that the chain is not a very accurate measuring instrument. Nevertheless, it will be seen later that it is adequate for the majority of chain surveys.

The advantages of the chain are that it is robust, vehicles may drive over it and the chain is rarely damaged. Due to its own weight it will, when correctly used, lie at ground level in fields of stubble, long grass and long weeds. On the other hand, it may damage plants if dragged through gardens and it cannot lie in a flat plane if the ground is very broken as on disused industrial sites. It should also be noted that some links may be bent by heavy vehicles passing over them, and mud, snow and ice may lodge between the small oval rings. The effect of any of these is to shorten the length of the chain, hence the surveyor must beware of these possibilities and check as necessary.

4.3.2   The tape

The *steel tape* is a continuous ribbon of steel, usually fixed into and carried in a steel, leather or plastic case. The tape is made in a variety of lengths, the most suitable for chain survey being 20 m, although this obviously depends upon the unit of measurement in use. Steel tapes should be used where it is not desirable to use a chain, e.g. where the chain might damage crops, or across a succession of garden fences. Steel tapes rust easily, hence they should be wiped dry and clean after use. They also kink easily and this frequently leads to their breaking. Even a bicycle passing over an untensioned steel tape lying flat may break the tape.

The *synthetic tape*, similar in shape and size to the steel tape, is manufactured from strands of glass fibre coated in PVC. It is used in chain survey to take offsets, the most suitable length being 10 m. Synthetic tapes are susceptible to length distortion, excessive tension causes them to stretch. Synthetic tapes should be used in place of the chain where it is necessary to take measurements in the vicinity of electric fences and railway lines. In such circumstances, the surveyor should ensure that the tape is dry.

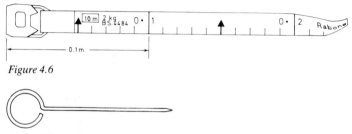

Figure 4.6

Figure 4.7

All tapes in chain survey are normally made with the measurement commencing from the outside of the ring, see *Figure 4.6*.

4.3.3 Arrows

The *chaining arrow* is a steel wire pin, roughly 0.35 m in length, as shown in *Figure 4.7*, used to mark the end of a chain or tape length laid down. Arrows are also used to record the number of chain or tape lengths laid down when measuring a line. One of the commonest blunders when measuring long lines is to forget exactly how many chain or tape lengths have been laid down. This may be overcome by always using a standard number of arrows, say five or ten.

The typical 'chaining' of a line requires a team of two, a surveyor and an assistant, the latter pulling the chain or tape forward while the surveyor 'drives' him from the rear. The surveyor lines in the assistant at the far end of the chain or tape, ensuring that he is holding the forward end correctly on the line which is being measured, then the assistant places an arrow vertically in the ground at the 20 m mark, or alternatively he makes a chalk mark, close to which he lays the arrow on the surface. When the chain is pulled forward again to lay down the next length, the surveyor walks forward and picks up the arrow. Since he starts at the beginning of the line with no arrows and the assistant starts with, say, ten arrows, then the number of arrows in the surveyor's hand will always indicate how many chain lengths have been laid in that line. (When the next line is started, the surveyor transfers all his arrows to the assistant again, so the assistant always starts a new line with the full quota of arrows.)

At the end of the chain line, or when measuring to any required point along the line, the value on the particular chain length is read and to it is added the distance indicated by the number of chain lengths shown by a count of the arrows in the surveyor's hand (often hung on the little finger). From time to time the surveyor should check the number of arrows held by both to ensure that none have been lost, since a lost arrow may cause incorrect recording of distance measured.

4.3.4 Ranging rods or poles

These are one- or two-piece poles of wood or metal, pointed at one end and made in various lengths, but typically 2 m (*Figure 4.8*). They are painted in bands of 0.2 or 0.5 m width, alternately red and/or black and white. They are useful for marking points on lines and the ends of lines to be measured. The preferred form is the 2 m rod with 0.2 m banding, since in addition to its main function this type is useful for measuring short offsets between the chain line and detail. The banding is an aid to visibility at long sights, but it may be better to attach a flag on very long sights. On hard surfaces, a tripod-form ranging rod support must be used.

4.3.5 The optical square

Several types of optical square are made, the best form having two pentagonal prisms mounted one above the other in a metal or plastic housing (*Figures 4.9 and 4.10*). The instrument is used for establishing and checking alignments and for 'raising' offsets.

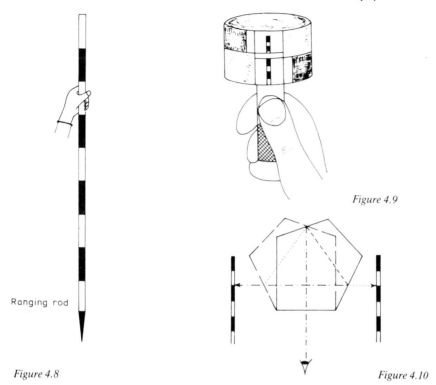

Figure 4.9

Ranging rod

Figure 4.8

Figure 4.10

To fix a point on a survey line, without sighting from one end of the line, the surveyor holds the instrument in front of his eye and walks across the line and at right angles to it. If a ranging rod has been placed at each end of the line, then when the instrument is exactly on the line each prism will show an image of one of the rods, and the images will be in line vertically.

To *raise an offset*, i.e. to drop a perpendicular from a detail point to a chain line, one of the prisms only is used. The surveyor places himself on the chain line, at the approximate position of the perpendicular and sights a rod at the far end of the chain line while simultaneously looking into the prism. Since the prism provides a view at right angles, the surveyor may move forward or backward until the image of the detail point appears in the prism and lines up with the directly viewed rod at the far end of the chain line.

4.3.6   The Abney level or clinometer

The *clinometer* is a hand-held instrument, used to observe the angle of slope of the ground along which a straight line is to be measured. The *Abney level* (*Figure 4.11*), the most popular type of clinometer, comprises a rectangular sighting tube, a graduated semi-circular vertical arc with vernier scale, a bubble tube, and a mirror. The mirror mounted in the tube enables the bubble to be observed in coincidence with the object being viewed (*Figure 4.12*).

The semi-circular arc is graduated in degrees, being read to 5 or 10 minutes of arc by estimation, using the vernier scale and a magnifying glass. The instrument is not suitable for use where slopes are excessively steep, or where high accuracy is demanded. Many Abney levels also have a scale of gradients on the graduated arc.

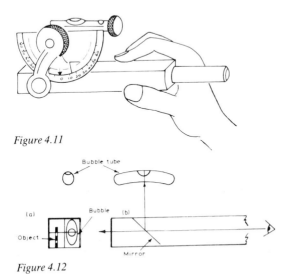

Figure 4.11

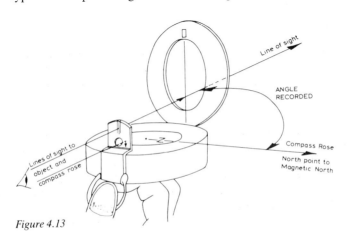

Figure 4.12

4.3.7 The prismatic compass

This is a small hand-held instrument used to obtain the bearing of one of the chain lines. This bearing may be used to orientate the survey. In Great Britain, however, the use of such an instrument is seldom necessary, since O.S. maps of the area of survey are generally available. The direction of North can usually be transferred from the O.S. map to the plotted plan with sufficient accuracy. Many varieties of prismatic compass are in use, a typical example being illustrated in *Figure 4.13*.

Figure 4.13

4.3.8   Ground marking equipment

In addition to arrows and ranging rods, the surveyor will probably require hammer, chisel, nails, wooden pegs, grease (wax) crayon, and chalk.

4.3.9   Booking equipment

Field book or booking sheets with A4 clip board, pen, ink, pencil, straightedge.

4.3.10   Plotting equipment

Drawing material, compasses, scales, straightedge, set squares, pencils, pens, inks, erasers, etc.

4.4   Standardization of chain and tapes

All linear measuring equipment must be checked from time to time against some *standard*. The standard might be a steel tape reserved for this particular purpose or, preferably, a standard distance laid down in, or adjacent to, the surveyor's place of work or office. A simple form of standard consists of two brass rivets or other marks exactly 20 m apart. The use of such a standard is described in Chapter 5.

4.5   Field procedure

Chain surveying consists, logically, of three distinct elements – the *lines*, the *measurements from the lines to the detail* and the *booking*. In much of the chaining process the three elements are concurrent, but in order to simplify matters for the reader, the elements are treated separately in the following sections.

4.5.1   Chain lines

4.5.1.1   *Reconnaissance and selection of lines*

On arrival on site, the surveyor walks over the entire area to be surveyed to familiarize himself with the general layout and in particular any problems it may present. Time spent on a thorough reconnaissance is never wasted. The survey lines and their layout are decided upon the basis of the principles set out below.

(i) The framework of lines must be primarily triangular. (A simple triangle around the site would be ideal, but this is rarely achieved.) There should be a *base line*, as long as practicable and preferably, but not necessarily, run through the middle of the area, the triangles being based on this base line. Sometimes a triangle must be set up, say to pick up detail, but it cannot be based on the base line. In such a case, the triangle may be erected on a side, or a part of a side, of another triangle, provided that the new triangle is 'tied back' to the base line by a suitable measured *check line*. When setting up a framework, triangles may overlap one another, and they may share common sides.

(ii) The triangles formed should be as few in number as possible, and they should be *well-conditioned* (i.e. as judged by eye, no angle of a triangle should appear to be less than 30° nor more than 120°).

(iii) Each triangle must be checked, therefore each triangle must be provided with one or more *check lines*.

(iv) Measurements to the detail must be kept short, see Section 4.5.2, therefore the chain lines should be positioned close to the detail but not so close as to cause difficulty in measuring the lines. The need to keep the chain line straight should have priority, however, and it is preferable to locate the lines along straight features or along points on the same alignment. For example, the faces of buildings or the tangent points of telegraph lines or the lines of straight kerbs or pavement edges are all suitable, provided that the resulting offset values do not become excessive.

(v) All detail which must be surveyed should be supplied by the methods described in Section 4.5.2. It is worth noting here that many inexperienced surveyors fail to position the lines to allow the detail at the corners of the site to be offsetted.

(vi) Obstacles to ranging and measuring should be avoided where possible.

(vii) Lines should be positioned over level ground if at all possible.

(viii) The total chainage involved in the survey should be close to the minimum required to achieve the above.

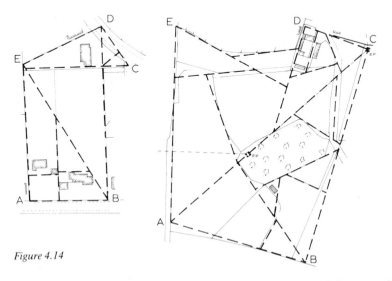

Figure 4.14

Figure 4.14 shows examples of frameworks selected for two different sites. It should be appreciated, however, that except on the simplest of sites, it is unlikely any two surveyors would select exactly the same arrangement of lines.

It will be noted that the chain lines start and finish either at the terminal points of other lines or at some point on another chain line. The terminal points of the former are known as (chain survey) *stations*, lettered A, B, C, etc., in *Figure 4.14*. The terminal point of a chain line that ties out at some point along another line is known as a *tie point*.

Stations may be marked in a variety of ways, depending upon the permanency required. On rock and similar surfaces, by a chiselled cross or painted mark, on softer surfaces a steel bar of 10 or 12 mm diameter may be driven, or a wooden peg about 40 mm square cross-section. Steel or wooden pegs can be surrounded in concrete and identification marks placed in the concrete when wet. Where marks are only required for the day – as typical on a small chain survey – then a ranging rod is simply stuck in the ground at the desired position. On hard ground the rod may be held above the mark by a tripod support.

Tie points are indicated by similar marks and are often described as *pickets*. These are normally established during the measuring process, e.g. in the larger survey diagram, the tentative position of all the lines would be decided at the reconnaissance stage, the positions of A, B, C, D and E being confirmed by inserting marks. Later, while measuring the base line AC, picket markers would be inserted to denote the intended tie positions. Where marks are to be concreted, this should be done before measurement commences.

4.5.1.2   Ranging the lines

It is generally important that there be no misalignment when measuring a line, otherwise any detail near the erroneous alignment and any chain line tying into it will also be in error. As a consequence, lines of more than one chain or tape length may need to be aligned by *tracing* or ranging before measurement. This is done by placing marks such as ranging rods or sticks

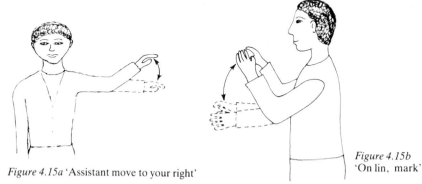

Figure 4.15a 'Assistant move to your right'

Figure 4.15b 'On lin, mark'

impaling pieces of paper at intervals along the line to be measured. Even the most experienced surveyor is liable to 'bend' a line by as much as 0.3 m if the alignment has not been fixed before measurement.

It is recommended that line ranging be carried out if the line exceeds 200 m length, or lies over very broken ground. For lines over 400 m length, the ranging should be by theodolite (see Chapter 5).

To *range a line*, the surveyor first marks the terminal points, or, if an end point is uncertain, places a mark on the line which will fix its direction. The surveyor then stands over the selected starting point and directs the assistant to proceed along the proposed line for about 100 paces, then turn and face the surveyor while holding a rod vertically. The surveyor signals the assistant to move the rod until it is on the correct alignment and on receiving the signal 'on line, mark' the assistant marks the position. The procedure is repeated at intervals of not more than 100 m until the end of the line is reached. It is desirable to place similar marks on the line in the vicinity of 'tie points'.

Ranging a line over a hill (or 'lifting a line') may cause problems. The easiest method is using the optical square, as in Section 4.3.5, but it should be done twice in opposite directions as a check. If no optical square is available, or the height differences are too great, the surveyor and assistant may repeatedly line each other in, as in *Figure 4.16*.

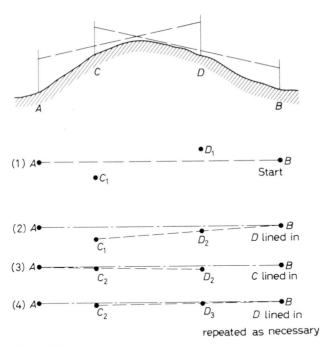

Figure 4.16

In the *Figure 4.16* A and B are the end stations. An assistant with the rod is placed at C, and another at D. C should be able to see a rod at B, and D able to see the rod at A. Neither C nor D is likely to be on the correct line. C now ranges D with himself and B, and in turn D ranges C in with himself and A. The result of this double movement will be that they both approach the true line AB. The alternate rangings are continued until no further movement is possible, then they are both on line. The successive positions have been shown on the figure as D1, C1, D2, C2, etc.

4.5.1.3   Line clearing

Lines are occasionally obstructed by vegetation which must be cleared to allow vision and chaining. Chain measurement is along the surface, so it may be necessary to cut bushes and very long grass. A bush-knife (matchet or panga) is most useful for this purpose, but a billhook will do. Care must be exercised with standing crops to avoid flattening wheat, maize, etc., or cutting cultivated trees and shrubs. On occasion it may be worth considering an alternative survey method, see Chapter 12.

4.5.1.4   Measuring the lines

The lines are chained, one at a time, measuring all detail and offsets before moving on to the next line. (The order in which the lines are measured is immaterial, except that unnecessary walking should be avoided. There is no particular merit in walking 5 km on a survey when a little thought would have allowed it to be done with only 3 km of walking!)

(i) Laying the chain
 The line having been ranged (if necessary) the chain is 'thrown out' or dragged in the direction it is desired to measure. The technique is for the surveyor to unwind a few links and, holding the handle(s) in one hand, throw the chain bundle forward along the line. He then stands on one handle while the assistant pulls the other handle forward along the line. The chain must be straightened by gently pulling and snaking it, and when it is straight the surveyor and assistant should inspect it for any tangled links which must be untangled.
 The outside of a handle is placed at the start of the line and retained in position by the surveyor (*Figure 4.17*) while the assistant at the other end holds the other handle and faces the surveyor. The assistant should hold

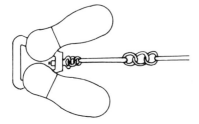

Figure 4.17

the arrows in the other hand or, preferably, in a small canvas bag or similar holder. The surveyor then directs the assistant on to the correct alignment and, when approximately on line, the assistant crouches while holding the chain with sufficient tension to keep it straight (*Figure 4.18*). The surveyor looks along the chain and 'through' the assistant's hand holding the chain to check correctness of the alignment, signalling to the assistant as necessary. Straightness and movement on to the correct alignment are achieved by the assistant snaking the chain and moving as needed slightly to one side or the other, then on the signal 'on line, mark' the assistant lowers his handle to the ground. The surveyor

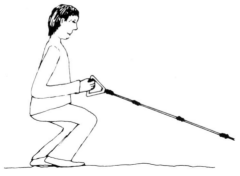

Figure 4.18

repeats the signal if the alignment is correct, otherwise the chain is realigned.

On receiving the second 'on line, mark' signal the assistant releases the handle so that the chain is no longer under tension. An arrow is then inserted in the ground against the chain handle, or on hard ground a chalk T mark is made and the arrow laid on the ground close to it (*Figure 4.19*). Note, however, that if the chain lies across hollows then some tension will be necessary. The surveyor should remain standing on the zero end of the chain until the far end has been adequately marked.

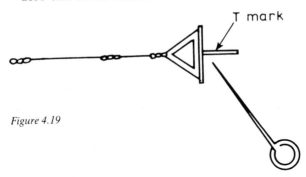

T mark

Figure 4.19

With the first length of any chain line laid, the assistant selects a 'back' object beyond the surveyor before either moves. This allows him to position himself on the approximate alignment for each succeeding chain length (*Figure 4.20*). When ready to move forward again the surveyor calls 'next chain' and the assistant drags the chain forward using the back object as a guide. When the assistant still has about a couple of metres to drag the chain the surveyor shouts 'check!', although a good assistant should be aware of the number of paces in a chain length and know when to stop. The process is repeated for laying the second and subsequent chain lengths.

The chain is read to the nearest 0.05 m at best, and if there is any doubt in the mind of the surveyor as to whether a length is, say, 72.35 or 72.40 m, then the lower value reading should be accepted. This is because in practice the length required is always less than the measured length, since very few lines are truly horizontal or on a level plane.

The chainage and offset values are read to 0.05 m, 0.1 m or 0.25 m, etc. The limit of measurement should be controlled by the plotting scale may be dictated by convention, by sheet size, by density of detail, by accuracy required and the probable need to achieve economy of effort, time and cost.

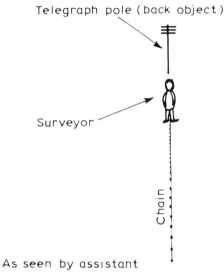

Telegraph pole (back object)

Surveyor

Chain

As seen by assistant

Figure 4.20

If, at any time, the site surveyor considers that the correct alignment is not being maintained, then it should be checked with the optical square.

The chain is usually dragged from line to line, being 'wrapped up' only at the end of the day's work. The 20 m chain may be wrapped up from one end in 'concertina' fashion, the bundle finally being tightly secured with a strap or sling passed through the handles, forming a wheatsheaf (see *Figure 4.2*).

If the steel tape is used as an alternative to the chain, then the technique is similar, care being taken to ensure it is free from twists along its length. Unlike the chain, tension is always applied when inserting the arrow or making a chalk mark at the 20 m end.

(ii) Slope correction

It will be evident that if the site is very undulating and has to be portrayed on a flat sheet of paper, then the ground measurements up and down the slopes must be reduced to a common datum, i.e. a horizontal plane.

This section deals with this problem, either by measuring horizontal distances as a series of horizontal steps or else indirectly by measuring the angles of slope of the lines being measured, then reducing the slope distances to their equivalent horizontal distances by calculation.

(a) Using step or drop chaining

This is best carried out measuring slopes downhill, in a series of horizontal steps, using the steel tape and locating a ground point vertically below some appropriate graduation reading on the tape ('plumbing' the measurement down) as in *Figure 4.21*.

The length of the step should not exceed 20 m in chain surveys, and the vertical drop at the end of the step should not exceed about 1.4 m (chest height). The assistant, at the forward end of the tape step, judges whether the tape is horizontal – this may be done reasonably accurately (within 2°) provided that the tape is not above the assistant's chest height. The steel tape is normally used because the chain, being much heavier, sags and so gives incorrect horizontal distances for the steps.

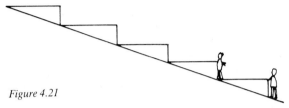

Figure 4.21

The ground point is 'plumbed' by dropping an arrow (special 'drop arrows' may be made) or a small object such as a pebble or a piece of chalk. The object should be dropped three times, to confirm the plumb point. The assistant should be re-aligned before each drop of the object and, for preference, plumbing should be from a whole number of metres. The procedure is repeated for each step.

Although it may seem rather crude the method is relatively fast and of sufficient accuracy for chain surveys.

(b) Using the Abney level

On long, even slopes it may be preferable to tape or chain on the surface, measure the angle of slope of the surface, then reduce the slope distance to the horizontal.

Initially, the surveyor stands close to the assistant and selects on him a mark which is at the same level as his own eye, or alternatively he may make an eye-level mark on a ranging rod. The surveyor then sends the assistant to one end of the slope, while remaining at the opposite end himself, and sights on to the selected mark with the Abney level (*Figure 4.22*). The observation may be checked by the two individuals

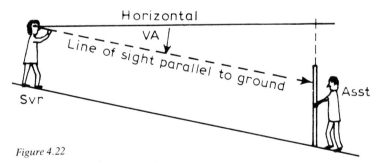

Figure 4.22

exchanging positions and repeating the slope measurement. Having obtained the slope distance and the vertical angle, the reduced horizontal distance is then calculated:

horizontal distance = slope distance × cos VA.

(iii) Obstacles to measurement

The methods described here have been applied by the authors and are considered to be the best. There are many alternatives but with some there is difficulty in finding a site large enough to set out the required geometric shapes, and in others there will be problems in adhering to the principles of surveying. Note that it is always best to use the simplest and most direct method – if it is possible to 'throw a tape across a river' then there is no point in using one of these methods, and where possible it is always better to measure straight through a building rather than around it.

The methods outlined below are typically used where a straight line XABY can be ranged visually, but it is not possible to measure the section AB directly. (A and B may be located on either side of a deep river or pond, a heavily trafficked road, or some similar obstacle to direct measurement.)

In some of these methods it is necessary to lay out one line at right angles to another (*raise a perpendicular* or *raise an offset*). If low accuracy is adequate (say, correct to 50 mm) then the methods of raising an offset may be appropriate, as described in Section 4.5.2. If higher accuracy is demanded then the right angle should be set out by theodolite (see Chapter 5) or by constructing a triangle with sides in the ratio 3:4:5 using one or two tapes.

The latter technique is illustrated in *Figure 4.23*.

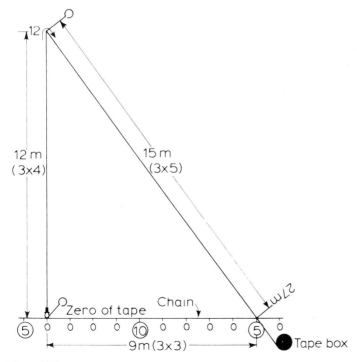

Figure 4.23

(a) Single 'A' method

In *Figure 4.24* it is required to find the distance AB along the line XY. Construct a well-conditioned triangle ABC, such that the line DE from the mid-point of the line AC to the mid-point of the line BC is clear of obstacles. Measure DE.

Now AB = 2 × DE, since the triangles ABC and DEC are similar.

Note: DE should be measured to twice the accuracy required for the measurement of the line XY.

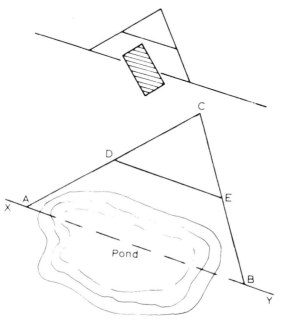

Figure 4.24

(b) Single 'X' method

Again (*Figure 4.25*) it is required to find the distance AB along the line XY. Construct two lines, AC and BD, clear of all obstacles, such that they intersect at their mid-points E. Measure CD.

Now AB = CD, since the triangles ABE and CDE are congruent.

(c) Inaccessible point method

In *Figure 4.26* it is required to find the distance AB along the line XY, Erect a perpendicular, AC, on the line XY. At the point C, construct a right angle BCD, such that D lies on the line XY.

Now AB = $(AC)^2/AD$, since the triangles ABC and ACD are similar.

The distance AC should be as long as possible, but it need be no longer than the unknown length AB. The construction of the right angles and the measurement of distances, for which the steel tape is used, should be to a fairly high standard of accuracy. Consider the following numerical examples:

If AC = 8.00 m, and AD = 4.00 m, then
    AB = 16.00 m,

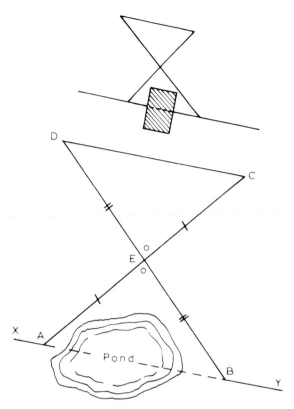

Figure 4.25

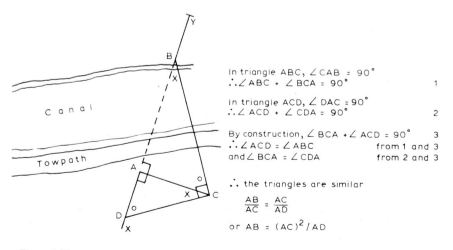

In triangle ABC, $\angle CAB = 90°$
$\therefore \angle ABC + \angle BCA = 90°$ 1

In triangle ACD, $\angle DAC = 90°$
$\therefore \angle ACD + \angle CDA = 90°$ 2

By construction, $\angle BCA + \angle ACD = 90°$ 3
$\therefore \angle ACD = \angle ABC$ from 1 and 3
and $\angle BCA = \angle CDA$ from 2 and 3

$\therefore$ the triangles are similar

$$\frac{AB}{AC} = \frac{AC}{AD}$$

or $AB = (AC)^2 / AD$

Figure 4.26

but

if AC = 8.01 m, and AD = 3.99 m, then
AB = 16.08 m.

Thus a 10 mm error in each measurement, or in the raising of the right angle, can cause an error of 80 mm.

(d) 14 02 10 method

This simple method (*Figure 4.27*) requires the use of a theodolite, but the reader who is not yet familiar with the instrument may nevertheless appreciate the method and put it into practice when he has covered Chapter 5 and has had the opportunity to handle a theodolite. The name of the method comes from the fact that a theodolite must be used to set out an angle of 14°02′10″.

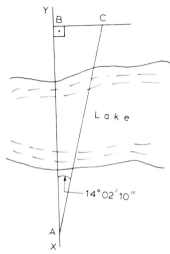

Tangent 14°02′10″ = 0.25 (i.e. ¼)

and tan 'A' = $\frac{BC}{AB}$

∴ $\frac{BC}{AB}$ = ¼, or AB = 4BC   *Figure 4.27*

It is required to find the distance AB along the line XY. At B, construct a right angle using a tape or the optical square. At A lay off (set out) an angle of 14°02′10″, such that it interesects the right angle raised from B at the point C, and measure BC.

Now AB = 4 × BC.

In practice, angles other than 14°02′10″ could be used, but the advantage of using this particular angle is that the perpendicular BC is relatively short as compared with the unknown distance, therefore errors in the setting out of this right angle will usually be minimal. The multiplication factor of 4 is also an easy number to multiply by, and to maintain the accuracy of the line XY it follows that the perpendicular BC must be measured to four times the accuracy required of the line itself, and again this is usually attainable.

(e) Measuring through a rectangular building
It is often necessary to measure through a building, but first it is usually essential to ascertain where the line actually enters and leaves the building. If this cannot be done, some re-arrangement of the lines will be needed. Ranging the line through a building may be done by (1) actually sighting through it, (2) sighting over the building, or (3) lifting the line over the building.

(1) Sighting through the building   Modern properties often have 'through' rooms with windows at either end. If these are suitably placed, then the line may be ranged using the optical square. The surveyor stands close to one of the windows at the outside of the building and, by moving parallel to the window, he may range the line through the two panes of glass.

(2) Sighting over the building   This is often possible if the building is single storey (e.g. a garage) or it lies in a hollow or valley. The surveyor stands at the start point of the line, identifies the far point, then places the assistant on line at the near face of the building. The procedure is then repeated from the far point of the line.

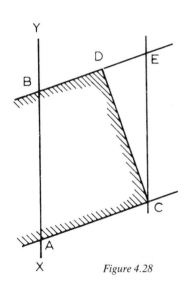

Figure 4.28

(3) Lifting a line over the building   This may be possible on flat roofed buildings with access, line points being marked on the roof just as they would be on the ground. Figure 4.28 shows a chain line XY passing over or through a rectangular building, the inaccessible distance being the distance AB and C and D being the corners of the building. To find the distance AB, measure AC and BD (assuming that AC is longer than BD) then measure out BE so BE = AC, and measure CE. Then AB = CE, since ABEC is a parallelogram.
   Should other detail obstruct the line DE and/or EC, then the length AB can be derived using Pythagoras' theorem.

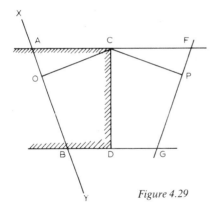

Figure 4.29

If the offset to C from line AB is required, then this may be achieved as shown in *Figure 4.29.*

Extend AC to F such that CF = AC,
extend BD to G such that DG = BD,
then AB = FG,
the distance AO = FP, a perpendicular having been dropped from C to FG at P,
and the offset OC = PC.

These methods may need to be modified if the wall faces are not mutually perpendicular.

4.5.2 Picking up the detail

Normally, detail in chain surveys is supplied by rectangular offsets from chain lines. However, it is not always feasible to use offsets and alternatives are sometimes needed. In areas of 'close' detail, in fact, considerable flexibility in the methods used is desirable. The methods used by the surveyor in practice are described here.

4.5.2.1   Offsets and running offsets

An offset has been defined as a short measurement taken (raised) at right angles from the chain line to the point of detail to be surveyed. Offsets are then said to be 'raised' and 'measured'.

Running offsets are offsets to two or more points of detail along the same perpendicular, as for example to both banks of a stream (*Figure 4.30*). Such offsets are called 'running' because the measurements are what are known in surveying as *running measurements*, that is to say successive measurements recorded along a line from some common point. *Figure 4.31* illustrates such a set of measurements, as do the measurements along a chain line itself. These forms of measurement are widely used in survey.

With experience it will be appreciated that running measurements, including running offsets, provide the quickest and most accurate method of measuring to a number of points on the same straight line.

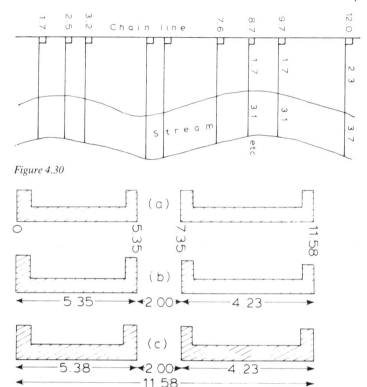

Figure 4.30

Figure 4.31 (a) Running measurements, (b) separate measurements, (c) separate and overall

(i) Recommended limitations

Offsets to points of detail such as corners, junctions and ends of detail should not normally exceed 8 m in length for scales of 1:500, 1:1000 and 1:1250, nor should they exceed 16 m for scales of 1:2000 and 1:2500.

Similarly, offsets to curving and indefinite detail such as meandering streams and hedges should not exceed 16 m in length for the larger scales or 20 m for the smaller scales.

These limitations are imposed by the need to ensure that the survey measurements are compatible with the maximum accuracy which may be demanded in the plotting, and the need to avoid long offsets which would require much additional care and time to maintain the possible accuracy.

(ii) Raising and measuring an offset

The assistant places the zero end of the tape at ground level on the point of detail to be surveyed. The surveyor holds the tape horizontal and taut at the approximate point on the chain where the offset is to be measured. Up to an offset length of 8 m, the site surveyor should have no problem in judging by eye the right angle required between the tape and the chain line correct to within 0.05 m. If the surveyor gently swings the taut tape from side to side in a horizontal plane like a slow moving pendulum it may assist his judgement of the right-angle and location of the correct point on the chain line.

Offsets up to 4 m in length are often more quickly raised and measured with a 2 m ranging rod rather than the tape.

When the line of sight between the chain and the detail point is obstructed, or the offset is more than 8 m in length, it is recommended that the right angle be raised with the optical square rather than by eye.

It will be shown later that some offsets are raised but not measured. Up to a distance of 8 m the surveyor may judge the location of the offset point on the chain quite easily without tape, pole or optical square. This is done by the surveyor standing and facing the point of detail to be surveyed with his toes close to the point on the chain at which it is anticipated the offset should be raised. It will then be found that the surveyor can, with ease, place the toe of his shoe on the chain at the required point.

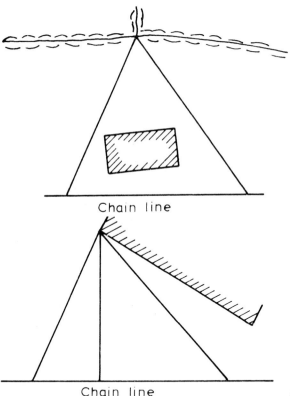

Chain line

Chain line

Figure 4.32

4.5.2.2   Ties

Ties are measurements taken from two or three different points on a chain line to a common point of detail which is to be surveyed.

Ties are used when it is considered that the length of a simple offset would be excessive, or when it is not possible to raise a right angle between the chain line and the point of detail (*Figure 4.32*). Though two measurements are usually sufficient over short distances, three may be preferred when long tape lengths are needed.

Recommended limitations
A point of detail to be fixed by ties should lie within 20 m of the chain line, unless a steel tape is used for the measurements. The length of the base of the triangle formed on the chain line should be the longest side length, so that a good 'cut' may be ensured when the point of detail is plotted, *Figure 4.33*.

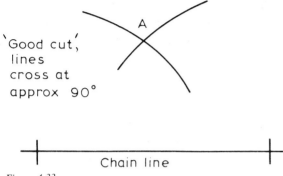

Figure 4.33

4.5.2.3   *Straights*

A *straight* is the extended alignment of any straight feature, including features such as the side of a building, a straight wall, etc. The intersection of a chain line and a straight is known as the *point of intercept*. The straight is little used in chain surveys, but it may be useful from time to time in areas of 'close' detail and in the unfortunate circumstance of the surveyor not having an assistant. The straight may be 'measured' or 'unmeasured' (*Figure 4.34*).

It must be appreciated that a straight cannot be plotted from a recorded straight measurement unless the other end of the straight is also surveyed by an offset, another straight, etc.

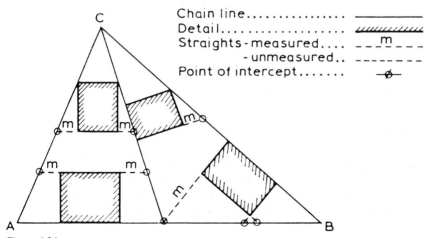

Figure 4.34

Recommended limitations

To minimize plotting errors which may arise due to possible misalignment of the chain lines, the angle at the point of intercept between a straight and the chain line should not be less than 40°.

Measured straights should not exceed 20 m in length, this meaning the distance along the straight between the point of intercept and the detail. Unmeasured straights may be of any length, provided that there is no doubt in the mind of the surveyor as to where the straight cuts the chain line.

It is often preferable that the chain line be arranged so that one end of a straight detail is close to the chain line, say within 20 m, but this will depend upon the length of the detail and the clarity with which the straight can be defined.

4.5.2.4   Plus measurements

Plus measurements are measurements made at right angles to a length of straight detail, often known as an *offsetted base*, which has been surveyed by offsets, straights, etc. Plus measurements are used to enable rectangular shaped buildings to be supplied (*Figure 4.35*).

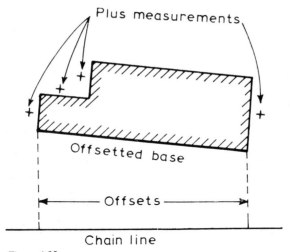

Figure 4.35

Recommended limitations

There should not be more than three plus measurements from either end of an offsetting base (*Figure 4.35*, left-hand side). The total length of the sum of the plus measurements at either end of an offsetted base should not exceed the length of the base, or 20 m, whichever is the least. The recommendations are intended to ensure that the plotting shall be up to an adequate standard.

4.5.3   Chain survey booking

Chain survey booking, may be carried out either on *booking sheets* or in a *field book*.

The field book can be used solely for chain survey or it may be used to record all the measurements required for some particular survey task. In the latter case, the field book might contain theodolite observations, chain survey measurements, and even the level bookings recorded in carrying out the same job.

The ideal booking sheet or field book for chain survey is A4 sized and of a plain or 5 mm square ruled paper. Chain survey books as obtainable from stationers or suppliers of office materials are usually small octavo size (between A5 and A6) and characterized by either a single or two red lines 15 mm apart down the centre of the page. While suited to the beginner, in practice these will be found inadequate since the page size is restrictively small and the position of the red lines may not always be convenient. With the recommended type, the surveyor may rule any lines to suit his own convenience.

Methods of booking vary but most surveyors adhere to a small number of basic rules which enable any other surveyor to understand what has been recorded. This is useful for a number of reasons, one being that it is possible for someone else to plot the work, a desirable feature since it goes some way towards providing the independent check identified in the principles of survey. It also means that should the surveyor be unable to complete the work for any reason, then others can interpret what was intended, what has been completed and what is still outstanding. Again, the surveyor's immediate superior will be able to assess how the work is progressing and criticize or compliment as appropriate.

Clarity and accuracy of booking are obviously essential. These may be achieved by neatness, taking especial care with one's handwriting, numbering and possible emphasis of detail, i.e. printing the numbers and using a slightly broader gauge of line to represent the detail. It is preferable to book in ink, unless the weather is inclement, a fountain pen being better than a ball-point pen for forming numbers and for line work. The surveyor must learn to do the bookings once only, since hand copying will lead to errors and considerably increase the time taken on the job, a satisfactory presentation and speed will come with practice. Mistakes inevitably occur, but no harm is done if these are carefully corrected – incorrect figures should be cancelled by drawing a single line through them, and revised values printed above or below the original figures, while cancelled detail should be indicated by small crosses on the cancelled lines.

The figures written on a page of field bookings may be written in different directions, according to the direction of measurement. Should there be any possibility of ambiguity as to the way up figures should be read then they should be underlined.

To the plotter, the field bookings are often the only indication he has of the site, thus not only the figures but also the representation of detail must be correct. Any symbols used must be understood by the plotter, and generally the symbol he draws on the map or plan is simply a neater representation of that used by the surveyor.

Explanatory notes may be added to assist the plotter and the numbering of pages is helpful and also provides a check as to whether any loose sheets may be missing.

It is unreal to expect the surveyor to be able to keep the pages of a field book clean, since chains and tapes become dirty and there is no way of preventing some of this dirt from being transferred to the book unless an extra (generally uneconomic) assistant is employed.

A sketch, an outdated plan or a smaller-scale plan attached to the field book will be found useful at the reconnaiserance stage, and the proposed stations and chain lines may be drawn on this. The first page of the field notes should include relevant information such as the name of the site or the job, the address, the name of the client if appropriate, the surveyor's name, the date, etc.

A chain line is usually represented by two parallel lines about 15 mm apart. The chainage figures are entered between these parallel lines, commencing at the *bottom* of the column, the distances to be entered being running measurements from the commencement of the line. The readings should be entered sequentially, and so spaced as to allow any entry to be amended or an additional entry inserted and no attempt should be made to observe scale. The length of the line is entered in the column on completion of the line, and it is emphasized either by circling it or 'topping and tailing' it, that is, entering the length sideways on between parallel lines in the direction of chainage, as in *Figure 4.36*. If plain paper is used for booking, guide lines in pencil to represent the chainage column are a useful aid, and they may be overruled in ink on the completion of each chain line.

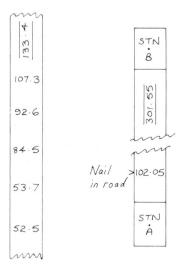

Figure 4.36

A *chain survey station* is represented by a square box at the end of the chainage column, the station letter (or number) being written within the box (*Figure 4.36*). An arrow head is used to denote a picket together with a brief description (see *Figure 4.36*).

Figure 4.37 shows how to represent tie points at the ends of a chain line 61.20 m in length, commencing at tie point 102.05 on the line from A to B and closing at the tie point 81.65 on the line 241.20. It should be noted that the tie point values are 'topped and tailed' for emphasis, and that the previously measured chain lines are repeated as single lines only, known as *skeleton lines*. Only the essential information about these skeleton lines is given, that is to say their length, page numbers and value of the picket (tie point) values being recorded and aligned as in the original chain lines. This duplication of information may not be necessary if the site is small and the plotting is to be carried out by the surveyor himself.

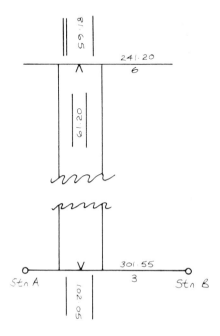

Figure 4.37

Occasionally, on his arrival on site to measure from or to a picket, the surveyor may find that the line originally planned is obstructed, perhaps by a heavy vehicle. If the obstruction cannot be moved then the surveyor may select an alternative chain line, thus either one or both of the planned tie points may have to be altered. Such a change may be booked as in *Figure 4.38* where 4.20 is the distance along the previous chain line between the original picket and the newly selected tie point.

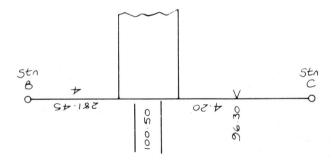

Figure 4.38

The value of 4.20 is shown in the booking because it provides a check for the plotter.

Offsets are booked as follows. If a point of detail lies to the left of the chain line (i.e. on the surveyor's left as he stands on the start point and looks towards the far end of the line) then the value of the offset to the detail is written to the left of and on the same line as, the appropriate chainage figure, but outside and as close as possible to the chainage column, (*Figure 4.39*). Where detail lies to the right of the chain line then similarly the offset value and the detail are shown to the right of the chainage column in the book.

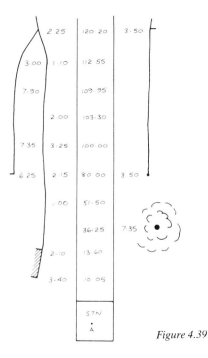

Figure 4.39

To avoid ambiguity in linking the value and its detail, the offset value is written close to the chainage column, and the detail is drawn close to the offset value. Mistakes in interpretation are less likely if adequate spacing is allowed between the chainage values, and some surveyors find squared paper useful for showing the relative alignments.

Running offsets are also depicted in *Figure 4.39*, and this illustrates the need to ensure that each offset value is shown close to the point of detail to which it refers.

All detail is assumed to be straight between consecutive offsets to the same detail, however much the freehand line drawn to represent the detail may meander, *unless* the surveyor states otherwise in the field bookings. In other words, the bookings are diagrammatic only. When taking offsets to curving detail, the usual practice is for the surveyor to raise and measure them at such intervals that for all practical purposes the curve between the offsets is assumed to be a straight line.

In urban areas it is common for a number of buildings to lie on the same alignment. Where such an alignment of buildings is to be surveyed by chaining, then it is not necessary to raise and measure offsets to every building corner on the alignment, as this practice is time-consuming and may lead to poor representation of the detail on the plotted plan. Every survey measure and every plotted point will have some accidental error, so that if such a row of buildings is each surveyed separately then, while each building should be correct on the plotted plan, the final alignment as a whole will not look precisely correct. This may be overcome by raising and measuring the corners at the ends of the alignment but only *raising* the others (*Figure 4.40*).

Note that the letters A and B circled on the example bookings indicate to the plotter that the detail is straight between the offsets at A and B.

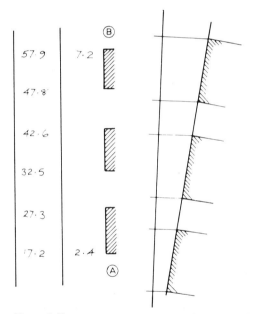

Figure 4.40

Ties may be booked as pecked lines with the measured lengths placed within the pecked lines, e.g., ____8.40____ (*Figure 4.41*). Similarly, an *unmeasured straight* is booked as a pecked line between the detail and the point of intercept, with the chainage value at the point of intercept circled.

A *measured straight* is booked also as a pecked line, but with its measured length circled and placed within the pecked line.

These arrangements of circles are used to avoid any possible confusion between offsets, ties and straights, since in practice it is possible for a single chainage value to refer to all three (*Figure 4.41*).

Plus measurements may be booked as in *Figure 4.42*. Note that the plus sign should always be entered nearest the point of detail at which the plus measurement is taken.

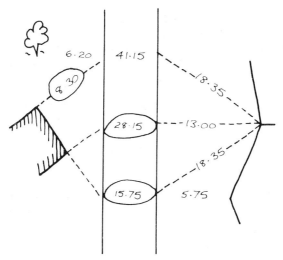

Figure 4.41

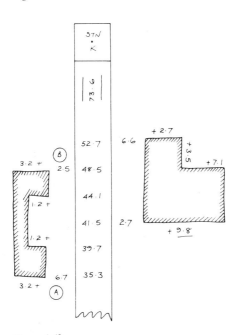

Figure 4.42

In chain survey it will often be found that detail connects between chain lines, and if the surveyor is not going to carry out the plotting then it is essential, particularly in larger surveys, that this detail be cross-referenced.

Where a point of detail supplied from one chain line connects through to a point of detail supplied from a later chain line, then the former point may be indicated by adding an arrowhead to the line of detail in the field bookings (*Figure 4.43*). The latter point is cross-referenced by the use of a skeleton line giving its length and page number, and repeating the former point's chainage and offset values (*Figure 4.44*).

To indicate where a chain line runs along the line of a kerb, or the face of a building, wall, fence, etc., the line representing the detail should be drawn in a much broader gauge of line where it is contiguous with the chain line (*Figure 4.45*).

When a chain line crosses the detail, the chainage value of the 'cut' should be recorded only if there is a bend, corner or junction in the detail at this particular point. The reason for this is that any small errors in alignment or measurement may cause a bend to appear in a straight feature when it is plotted. Examples of the booking of detail cutting the chain line are shown in *Figure 4.46*.

The above rules cover most contingencies on site, if it is considered that at any time there may be some doubt in the mind of the plotter, then add

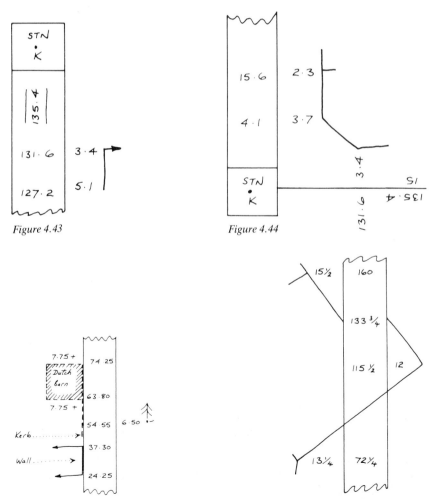

Figure 4.43

Figure 4.44

Figure 4.45

Figure 4.46

an explanatory note. Always when booking, consider the plotting; remember that ideally there should be no need for the plotter to refer back to the surveyor. The bookings should be complete, clear and accurate.

4.6   Office procedure and plotting

Chain survey *plotting* requires no great skill, although the draughtsman must be accurate and meticulous. Graphic communication skills are of importance in the presentation of the finished work, however, and these are considered here.

4.6.1   Preliminary considerations

4.6.1.1   *Choice of drawing material*

The choice of material for the plan or map will depend upon the purpose and use of the plotted survey plan. The material may need to be durable, to be transparent or translucent for purposes of copying and tracing and it may need dimensional stability. The need to take ink and pencil is evident and on occasion it may be necessary for the material to take colour washes. Also, the ability to withstand frequent erasures is generally desirable.

There are three materials in common use – cartridge paper, tracing paper and polyester-based translucent films.

If *cartridge paper* is used, then the heavier and higher quality grades are the more suitable, but a sample should be tested for ability to take ink and colour washes and to withstand erasing. Cartridge paper is not dimensionally stable, and perfect erasure of inks is not possible.

Tracing paper is probably the most commonly used material and again the heavier grades are better, but it is not dimensionally stable and becomes brittle with age.

Polyester film is strong, durable and dimensionally stable, probably the best material to use but it is more expensive than either cartridge or tracing paper. There is a slight tendency for ink to spread on film and pencils must either be of a hard grade or have the more recently introduced plastic-based leads. Both ink and pencil can be erased.

4.6.1.2   *Sheet size*

Sheet size is obviously linked to the chosen scale and the area to be surveyed and on a large survey it may be desirable to have two or more sheets. If copies are required then the reprographic facilities available must be considered. Sheet size will also depend upon the amount of marginal and border information required to be shown, for example a title, a north point, a legend, road destinations, etc.

4.6.1.3   *Layout on the drawing material*

Having decided on the material and the sheet size to be used, there are three other matters to be considered before plotting can be commenced. These are the orientation and position of the survey on the sheet and the border and marginal information to be provided. When plotted and penned, the drawing should present a balanced appearance pleasing to the eye and much of this will be achieved by the quality of the line work and the lettering. Often, North is placed towards the top of the sheet, but this is not essential.

4.6.2   Plotting the framework – line plotting

The equipment required includes a scale, steel straight edge (metric if graduated), pair of compasses, beam compass, pencil, pencil sharpener and eraser. The pencil should be one which will produce a fine clean line, of even width and density, which will not smudge but may be erased without undue damage to the drawing surface. This means a good quality 'hard' pencil sharpened to a fine point.

Plotting should commence with the drawing of the base line, in a suitable position on the drawing material and to the appropriate scale, together with any relevant tie points. Stations and tie points may be lightly encircled, with their values written adjacent to the line. It should be remembered that all this may have to be erased later.

The largest and best-conditioned triangle tied to the base should be plotted next by using the compasses to describe intersecting scaled arcs of the recorded distances. The chain lines are then drawn-in with the straight edge and their lengths checked with the scale. It is most important to ensure that the drawn lines pass exactly through the plotted points – beginners often have initial difficulty in doing this. Next, plot and check any lengths within the plotted triangle and add the relevant chain lengths and tie-point values. The remaining chain lines are plotted in a similar manner.

For checking the lengths of lines, the limiting tolerance usually accepted is 1:500, though occasionally 1:1000 is required. Further, no line should be rejected if the difference between the measured and the plotted lengths is less than three times the recommended limits of measurement of the scale of plotting. Thus, no lines with the following errors should be rejected:

0.15 m at scale 1:500
0.30 m at scale 1:1000
0.75 m at scale 1:2500

Should a line fail to plot within these limits, and the misclosure is large, then the line in error and/or its tie points must be checked and re-measured as necessary. On the other hand, if the misclosure in a line is small, then a judicious adjustment of the surrounding lines may enable it to be fitted into the plot within the above limits of tolerance. That is to say, the network holding the line in error may be expanded or contracted so as to bring all the lines within the limits of tolerance.

Where a line has an acceptable misclosure, then the plotted value of any tie point on that line is to be equated proportionally.

4.6.3   Plotting the detail

Plot all the detail from one chain line, complete and draw in, then move on to the next line, and repeat until the detail plotting is finished. Plot the chainage values as short ticks. The offsets are raised using a set square and straight edge. The scale is used to plot the offset values. As an alternative to this, offset scales are faster if available. A pair of compasses is required to plot the ties. The straights, whether measured or unmeasured, cannot be plotted until the far end of the straight has been located on some other chain line.

Detail points are then connected as indicated in the field bookings, using the straight edge. Plus measurements may then be plotted using the set square, straight edge and scale in a manner similar to the plotting of offsets.

Where a line has an acceptable misclosure, the chainage values as recorded should be increased or decreased proportionally.

Where the acceptable misclosure is at the minimum value, such as 0.15 m at 1:500 scale, 0.3 m at 1:1000 scale, etc., then the misclosure may be 'lost' at the terminal points of the line without adjusting intermediate chainages.

4.6.4 Completing the plan

The penning-in of detail can be carried out as plotting proceeds, possibly line by line but preferably on completion of the pencil work.

When all detail and symbols have been inserted, add any necessary notes, such as direction of roads 'From Leicester . . . To Birmingham', names of buildings and areas, etc.

In positioning the name of a feature on a map or plan there should be no possibility of confusion as to which feature the name refers to, nor should the name obscure any other part of the drawing.

Names identifying linear features such as roads and rivers should be aligned within or alongside the feature, spaced so as to indicate the feature's extent. Names identifying areas, such as field and wood names, should be placed centrally within the area and with sufficient spread to give an indication of the extent of the area.

A simple North point must always be drawn, and a drawn scale should be placed at the bottom of the sheet. It is traditional that the drawn scale should be as long as the longest line in the survey, but this is not essential. The drawn scale must have its description clearly marked. 'Scale – 1:500, metres.'

The usual title, firm's name, job number, and the names of persons responsible for surveying, plotting and tracing the work, date, etc., should be added.

Colour wash should now be applied if required, a small amount does enhance the finished plan. However, it is not satisfactory if copies of the plan are to be reproduced by the cheaper reprographic processes. Dry transfer hatchings, shadings and textures are possible alternatives.

Finally, remember the plan has to communicate adequately and accurately its contents to the user.

Chapter 5

Control

5.1 Introduction

Control was referred to briefly in Chapter 1 as being the principle of tying measurements together in such a way as to produce a survey which is accurate in proportion and scale. In a small chain survey, the chain lines are measured to a higher standard of accuracy than the detail measurements, thus the chain survey framework controls the detail measurements.

Where a chain survey is very extensive, the standard of accuracy of ordinary chaining may be inadequate to keep the whole survey accurate in proportion and scale, and it then becomes necessary to provide a higher accuracy framework to control the chain survey. In this case, there are three levels of accuracy, the highest being the control frame, then the chain lines and finally the detail measurements. The *control* would be plotted first, then the lower accuracy chain lines fitted to the control and then the detail plotted from the chain lines. This demonstrates the survey maxim of *working from the whole to the part*.

In a very large task, there may be successive levels of control. Thus the Ordnance Survey's primary triangulation of the United Kingdom, measured to the highest standards, is described as 1st order control, the infill between these is 2nd order control, and ultimately fitted to this is the 3rd order work which actually controls the detail measurements.

Control frameworks, then, are measured to a higher standard than can be attained in chain survey and they are likely to involve both linear and angular measurement. Control may consist of a single base line, or a triangulation scheme from a base line, or a trilateration network, or one or more traverses.

Generally the most economical method in terms of time and money is the *traverse*. For a large area, a network of traverses may be needed, perhaps tied to a base and a small triangulation scheme.

This chapter considers the methods and equipment used in linear and angular measurements for the provision of control.

5.2   Linear measurement

Traditional 'direct' linear measuring equipment includes the chain and tapes described in the previous chapter, together with the steel band, a form of steel tape. Distance measurement by optical and electro-magnetic distance measuring equipment is covered in Chapters 12 and 13.

The higher standards of accuracy demanded of control measurements using the traditional linear measuring equipment are achieved by the use of particular techniques and procedures. These include

Measuring each distance at least twice, possibly in one direction in metres and in the other direction in feet. This can improve accuracy and provide an independent check.

Ensuring that the steel band or tape is correctly aligned at all times during the actual measurement. Failure to do this will introduce a cumulative error.

Making the appropriate corrections when measuring along sloping ground.

Applying the correct tension to the band or tape when marking each band or tape length.

Applying corrections to the measured length to allow for the effect of temperature changes causing expansion or contraction of the band or tape.

Making a suitably fine mark to indicate the end of each band or tape length.

Standardizing the band or tape before starting the work.

In the measurement of lines by traditional methods, the band or tape is laid on the ground surface as described in Chapter 4, this method being known as *surface taping*. Alternatively, the band or tape may be suspended in *bays*, clear of the ground, supported by tripods, wooden posts, poles or pegs, this method being known as *catenary taping*. (The *catenary* is the curve formed by a chain, tape or rope of uniform section and density, suspended between two points – a bay.)

Surface taping can achieve accuracies of up to 1 in 10 000 with relative ease on the majority of survey sites, and this is adequate for most tasks.

Accuracies greater than 1 in 10 000 are needed occasionally, as for example in some base lines and in setting out some prefabricated structures. In the past, the customary method for obtaining such higher accuracies was catenary taping, since this eliminated the effects of surface irregularities, reduced the uncertainty in the determination of the tape temperature, and allowed greater flexibility in the positioning of the line to be measured. Today, higher accuracy measurement is more usually carried out using electro-magnetic distance measuring equipment (EDM).

Where EDM equipment is not available, the odd distance can be found by the sub-base technique described in Section 12.2. Alternatively, surface taping can achieve accuracies greater than 1 in 10 000, provided the conditions are suitable, i.e. long, even ground slopes, smooth ground surface, shady conditions, etc.

5.2.1   Equipment

5.2.1.1   The steel tape

This is described in Chapter 4.

5.2.1.2   The steel band

This is similar to the steel tape, but is carried on a cruciform frame rather than in a box. It is generally preferred to the tape when higher standards of accuracy are demanded, although it is not itself more accurate. The features which make it more suitable for higher accuracy work include

(i) It is generally of greater length than the tape.
(ii) It is narrower and thinner in section, hence its weight per unit length is less than for a tape. This makes it more suitable when measuring in catenary, since sag and sag correction are reduced.
(iii) The zero marks are etched at some 150 mm in from the ends of the band, rather than at the outside of the end rings, thus measurements are not affected by distortion of the end rings.
(iv) The end rings may be used to fix detachable handles, similar to chain handles, or leather thongs.
(v) The band is readily detached from its carrying frame, allowing easy attachment of tensioning devices.

Figure 5.1 shows a steel band in its winding frame.

Figure 5.1

5.2.1.3   Ancillary equipment

The following items may be required:

(i) Thermometer. Used to record the temperature of the band.
(ii) Spring balance. Used to ensure that the correct tension is applied to the band.
(iii) Tape grip (Littlejohn roller tape grip). Allows the spring balance to be attached to the band at any point in its length. This is often necessary when tension must be applied to a short length of band.

(iv) Theodolite. This instrument, described in Section 5.3, is used to observe the angle of slope of the ground and for accurate ranging of the line which is to be measured. As an alternative, a level and staff (see Chapter 8) may be used to obtain the height difference between points on a line, from which the reduced horizontal length of the line between the points may be calculated.

(v) Arrows, as described in Chapter 4. These are used to indicate the end of each steel band or tape length laid down. Where high accuracy is required, then instead a fine mark must be drawn, cut or scratched on a smooth surface. On a rough surface, a fine·surface for marking may be obtained by pressing out a thin layer of plasticine over the surface then making the mark in the plasticine with a penknife. Mud may be a substitute in rural areas.

(vi) Other equipment in Chapter 4. This may include optical square, ranging rods, pegs, hammers, nails, booking sheets, etc.

5.2.2 Corrections and errors in measurement

It is usually necessary to apply *corrections* to measurements made by direct linear methods, the particular corrections to be made depending upon the standards of accuracy demanded of the work. An accuracy of 1:2000 can be achieved simply by using a steel tape, provided corrections for slope are applied, the tape is in good order and the standard procedures are followed with care.

An accuracy of 1:5000 requires a steel band in good order, with corrections for slope applied, the correct tension (specified by the maker) applied, and corrections applied for variation of temperature from standard. An accuracy of 1:10 000 would, in addition, require that the band be *standardized* (checked for length) and that all lines were measured twice. To achieve better than this, then the line positions must be such that the ground slopes are long, even and smooth, and the work is not carried out in direct sun.

5.2.2.1 *Standardization*

If a steel band or tape is longer or shorter than its *nominal length* (the length marked on it) a cumulative error will result when it is used to measure lines. Bands and tapes should be checked against some standard from time to time, then appropriate corrections can be applied to the measured distances. The *standard* may be a steel band reserved solely for this purpose and the determination of the correction is termed *standardization*. Standardization may be used to find the true length of the band, or to find a *standardization factor* such that

True line length = measured line length × standardization correction
factor

The standardization correction factor is obtained by setting out a horizontal bay, slightly shorter than the length of the band to be standardized, then measuring the bay length with the standard band and the field (work) band, in each case applying the correct tension for the particular band and allowing for temperature differences. Then

$$\text{Standardization correction factor} = \frac{\text{Length recorded on standard band at } x \, °C}{\text{Length recorded on field band at } x \, °C}$$

An answer to five decimal places is usually adequate, and the result cannot be expected to be more accurate than that attained with the standard band. A new band is often used as a standard and manufacturers state generally that this may be accurate only to ±2.5 mm/30 m length. Some manufacturers issue a certificate for a band if a higher accuracy is required.

Unless a band is damaged, or is very old, standardization is only necessary when accuracies greater than 1:5000 are demanded.

5.2.2.2   Temperature

Temperature may be observed by placing a thermometer on the ground just before measurement, or twice a day in sun and shade, or again for every line measured, subject to the accuracy required.

For an accuracy of 1:5000, the band temperature should be known to within ±5°C. For an accuracy of 1:10000, the temperature should be known to within ±3°C.

Note that when a band lies on the surface of the ground in the sun, then the band temperature may be 5°C or more greater than the ground surface.

Manufacturers state that a band reads correctly at a particular temperature under a particular tension, this particular (*standard*) temperature usually being 20°C. The coefficient of linear expansion for steel bands and tapes is generally 0.000011/°C, and the correction for temperature variation is

Line correction = Temperature variation from standard × coefficient of linear expansion × length of line

or

$$c = \Delta t \times 0.000011 \times l_m$$
$$= 0.000011 \, (t_f - t_s) \, l_m$$

in which t_f and t_s are the *standard* and *field* temperatures respectively.

If the corrected length = l_c, then

$$l_c = l_m + 0.000011 \, (t_f - t_s) \, l_m$$
$$= l_m \, (1 + 0.000011(t_f - t_s))$$

If temperature correction factor = f, then

$$l_c = l_m \times f$$

and

$$f = 1 + 0.000011(t_s - t_f)$$

As a rough indication, a 100 m band will be 22 mm shorter when the temperature is 0°C than when the band is at the standard temperature of 20°C. Occasionally surveyors combine the above two corrections to produce the temperature at which the field band reads its true length.

5.2.2.3   Tension

Field tests have shown that failure to apply the correct tension with a spring balance may result in readings which are in error by as much as 15 mm/30 m, i.e. 1:2000. This indicates that where an accuracy of greater than 1:2000 is required a spring balance should be used to apply the correct tension to the band. The balance is attached to the forward end of the band for tensioning, using a tape grip if necessary. The tension to be used is stated on the band, e.g. 5 kgf or 50 N, or 7 kgf or 70 N.

Errors which arise from a failure to use the standard tension may be corrected by re-standardizing the field band at the incorrect tension. Alternatively,

Correction to measured length $= \pm\, l\, (T_f - T_s)/AE,$

where l = measured length, T_f = tension used in the field, T_s = standard tension, A cross-sectional area of the band, and E = Young's Modulus of Elasticity. The units must be compatible, thus l and the correction in metres, T in Newtons, A in mm^2 and E in Newtons per square millimetre.

If an accuracy of 1:10 000 is required, then the tension should be correct within $\pm$ 1.5 kgf or 15 N. Spring balances should be checked against one another occasionally.

5.2.2.4   Bad alignment

Where band lengths are aligned by eye, a common practice over short distances, lines may be misaligned by as much as 100 mm/100 m, a bow of 300 mm in a 300 m line. This has the effect of increasing the apparent line length by about 0.4 mm/100 m, an amount which may be ignored in all except the most accurate work. (It must be re-emphasized here that although chain survey does not require the high accuracy of control measurements, misalignment of lines should be avoided in chain survey since offsets or lines tied to bowed lines may be subject to serious errors.)

Poor alignment or bowing may also occur along each band length, that is to say individual band lengths may not be laid straight. Causes may be a strong wind blowing across the line, the line passing through a fence post or blocked by a car wheel or vegetation, these causing the line to be bent round the obstruction. All these forms of misalignment should be guarded against. Note, however, that the odd bowing of a band may be ignored on occasion, a bow of 50 mm in a 30 m band length along a 100 m line producing an error of less than 0.2 mm in the total line length.

Vertical misalignment occurs when the line of sight of the angle measuring instrument is not parallel to the ground, as in *Figure 5.2*, where h is not equal to i. Surface irregularities, undulations, cause effects similar to horizontal misalignment, and all result in cumulative error.

5.2.2.5   Slope

All linear measurements for control should be reduced to the horizontal. Traditional linear measurements are made on the surface of the ground, unless measuring in catenary, and the angle of slope is read on a theodolite by observing a line parallel to the line measured with the band. This is similar to the use of the Abney level in chain survey, illustrated in *Figure 5.2*.

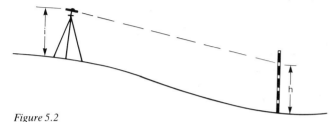

Figure 5.2

Horizontal distance = slope distance × cosine of the angle of slope, thus the *slope correction factor* is the cosine of the vertical angle. On occasion, as in Section 5.2.1, the difference in height of the two ends of the line is obtained, then the horizontal distance is calculated most simply by Pythagoras' theorem.

Although slope correction by cosine of the slope angle is easily applied, the accuracy of the slope correction itself depends upon the accuracy with which the slope angle is measured and the steeper the slope the more accurately the angle must be measured.

Where a measured length is required to be accurate to 1:5000 or 1:10 000, then the slope correction must be accurate to within about 2 mm/30 m for each measure of the line. To achieve this, the vertical angle (VA degrees) must be correct to within approximately 600/VA seconds of arc, thus on a slope of about 2° the angle must be measured to within 300" or 5' of arc. Similarly, a slope of about 10° must be measured to within 60" or 1' of arc.

In *Figure 5.2*, the distances h and i should be equal to within about 300/VA mm for every 100 m of line length. Thus for a 2° slope, h and i may differ by up to 150 mm/100 m, but on a 10° slope not more than 30 mm/100 m of line length.

A common problem is that of identifying where changes of slope actually occur and thus defining the separate slopes. No serious guidance can be given in this respect, however, only practice will develop some skill.

When taping over broken ground, it is permissible to suspend a short length of the band in catenary, typically up to 15 m or so in length. This depends upon the band weight per unit length, however, and the greater the weight the less must be the length of the suspended span.

5.2.2.6   Marking band ends, reading the band

Each time a band is laid down, errors may arise in marking the end of the band and in reading the graduations on the band. These may tend to compensate, but it is good practice to measure each line twice, particularly if an accuracy of 1:10 000 is required, and demand that the two measures agree within some standard such as 2 mm/30 m band or part band length.

5.2.2.7   Projection and scale

When tying linear measurements to a higher order of control, such as that of the Ordnance Survey, a *projection* or *scale correction* is necessary, as referred to in Section 2.3. Every reduced horizontal distance in the survey must be multiplied by the Ordnance Survey *local scale factor* to give its equivalent National Grid distance, Section 6.5.1 refers.

5.2.2.8   Altitude (height above sea level)

In higher accuracy work, measured lengths may need to be reduced to the equivalent distance at mean sea level, as in *Figure 5.3*.

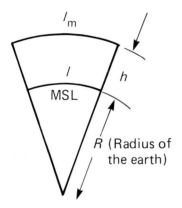

Figure 5.3

If l_m = length measured, l = length at mean sea level, and h = mean altitude of the measured line above mean sea level, then

$$l/R = l_m/(R + h)$$

or

$$l = l_m R/(R + h)$$

R may be taken as 6 370 000 m and a table of correction factors can be produced.

5.2.3   Booking and reduction of measured lengths

See Section 6.5.1.

5.3   Angular measurement and the theodolite

Angular measurement covers the measurement of angles in both a horizontal and a vertical plane. The traditional instrument for angle measurement, capable of providing an angular accuracy compatible with that of linear measurement, is the *theodolite*.

The theodolite is designed specifically for the measurement of horizontal and vertical angles in surveying and in construction works. Although the instrument's basic principles are simple, it is the most versatile of survey instruments, capable of performing a wide range of tasks. In addition to the measurement of angles, these include setting out lines and angles, levelling, optical distance measurement, plumbing tall buildings and deep shafts, and geographical position fixing from observations of the sun or stars, etc.

It can be an extremely accurate piece of equipment, some instruments being capable of being read to the nearest minute of arc, while others can be read to 0.1 of a second of arc.

5.3.1   Theodolite classifications

Theodolites are typically classed according to their reading accuracy, and each class has a generally accepted range of characteristics.

5.3.1.1   Precision theodolites

These may be read directly to better than one second of arc and they are typically used by national mapping and land survey organizations for geodetic surveying.

5.3.1.2   Universal theodolites

Also known as 'single second' instruments, these allow readings directly to 1″ of arc. They are occasionally used on site survey and engineering work, especially when extreme angular accuracy is needed, such as possibly on very long sights.

5.3.1.3   General purpose theodolites

Generally these read directly to about 20″, and they are fast and easy to use, ideal for general survey work. Often described as 'engineer's' theodolites.

5.3.1.4   Builder's theodolites

These instruments are of a comparatively low order of accuracy, reading direct to 1, 5 or 10′ of arc and by estimation to 30″ or 1′. They are usually rugged, simple to operate, and relatively cheap.

5.3.2   The basic construction of the theodolite

Figure 5.4 shows the basic construction of early instruments, first made about 400 years ago, and modern instruments are merely refined versions of this original concept.

A circular protractor, today called the *horizontal circle* or *lower plate*, was supported horizontally with its engraved surface lying uppermost. Above this was an *upper plate* with an index pointer, the plate being rotatable to allow the pointer to be laid against any graduation on the lower plate. The upper plate carried two vertical A-frames (now known as

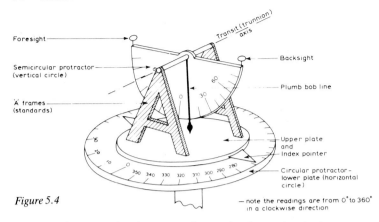

Figure 5.4

standards) supporting between them a horizontal axis, known as the *transit* or *trunnion axis*, at the centre of which a semicircular protractor was rigidly fixed at right angles to it. A weighted plumbline was also carried by the transit axis. (The modern protractor is a full circle and is termed the *vertical circle*). On its diameter the semicircle carried two sights, as on a rifle, which defined the line of sight or line of collimation. To read at the

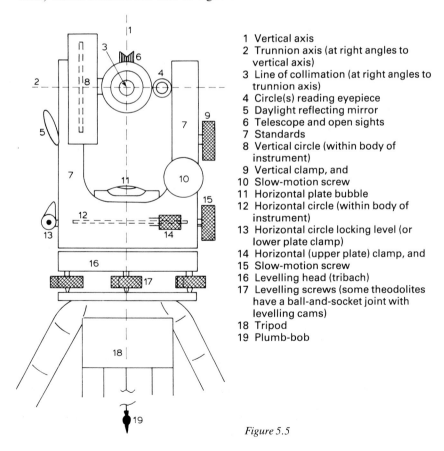

1 Vertical axis
2 Trunnion axis (at right angles to vertical axis)
3 Line of collimation (at right angles to trunnion axis)
4 Circle(s) reading eyepiece
5 Daylight reflecting mirror
6 Telescope and open sights
7 Standards
8 Vertical circle (within body of instrument)
9 Vertical clamp, and
10 Slow-motion screw
11 Horizontal plate bubble
12 Horizontal circle (within body of instrument)
13 Horizontal circle locking level (or lower plate clamp)
14 Horizontal (upper plate) clamp, and
15 Slow-motion screw
16 Levelling head (tribach)
17 Levelling screws (some theodolites have a ball-and-socket joint with levelling cams)
18 Tripod
19 Plumb-bob

Figure 5.5

instrument station a horizontal angle between two targets, the sights were aimed alternately at the left-hand and right-hand targets and the horizontal circle readings to them were noted, the difference between these readings giving the horizontal angle between the directions to the targets as measured at the instrument station. To read a vertical angle, the sights were aimed at a target and the angle of elevation or depression read off the vertical circle against the weighted plumbline.

Today theodolites have telescopes to improve the line of sight in distance and clarity of definition. They also have glass circles within the body of the instrument, replacing the original brass protractors, thus the circles may be graduated more finely and light can be reflected through the circles to show the markings with greater clarity.

Figure 5.5 shows the essential components of such a theodolite. Note that when moving a theodolite from its case to the tripod, it should be carried by the two standards rising from the upper plate surface. A new generation of theodolites is being developed by some manufacturers in which electronic digital reading systems are superseding the glass circles in common use today.

5.3.3   The tripod and tribrach

5.3.3.1   The tripod

The *tripod* is a device for supporting the instrument at about eye height over survey marks. The modern tripod has three framed legs, which may be rigid or telescopic. The former is more stable in use, the latter more adaptable, e.g. if the surveyor is very tall or very short, or on sloping ground.

The tripod must be set up with the feet of the legs about one metre apart and so that it is stable and at a convenient height for the observer. The tripod feet should be well pressed into the ground and on slopes, one leg should point uphill for stability. The legs should be oriented in such a way as to avoid the surveyor having to straddle a leg while observing. The top must be approximately horizontal and central over the survey mark. In hot climates it should be set up in a shady position, or a survey umbrella used for protection of the instrument. There should be no loose fittings. Some versions are equipped with a *centring rod* (like a non-supporting central fourth leg) which has an attached circular bubble and is used for centring the instrument over the ground mark.

In some jobs, it is necessary to place the theodolite on top of a wall or pillar and in these circumstances a special *base-plate* (or *wall plate* or *trivet*) is used instead of a tripod.

5.3.3.2   The tribrach or levelling head

The prime purpose of the *tribrach* (or levelling head) is to facilitate the levelling-up of the instrument, so that readings are taken in truly horizontal and vertical planes. The *levelling-up* operation involves making the vertical axis of the instrument properly vertical.

When travelling to and from survey sites, the theodolite is clamped securely in a carrying case and the tripod is carried separately by its carrying handle or strap. The tribrach usually forms the lower part of the theodolite, the instrument being clamped to the tripod by a captive bolt screwed into the tribrach when the instrument is set up, but on some theodolites the tribrach is part of the top of the tripod.

Where the tribrach forms part of the instrument, the captive bolt may be eased and the instrument slid over the top of the tripod to assist centring over the ground mark. With this form of construction, care must be taken to protect the tripod top from damage.

The tribrach includes *levelling screws* or *cams*, used to tilt the instrument in levelling up. The use of these is described below. In many instruments in which the tribrach is part of the theodolite, the tribrach is actually detachable. This permits other items of equipment (traverse targets, etc.) to be mounted in the tribrach on top of the tripod in place of the theodolite.

The theodolite in use must be centred over a ground mark, and traditionally this is done by reference to a *plumb-bob* and *string* suspended from the centre of the underside of the tribrach. The plumb-bob is generally carried in the instrument case or in a pouch attached to the tripod.

Centring the instrument is achieved by appropriate movement (laterally, or by extending or shortening) of one or more of the tripod legs, until the plumb-bob is exactly over the ground mark, with the final fine adjustment being made by unclamping the tribrach and sliding it across the top of the tripod head. On some few instruments, the upper part of the instrument may be slid over the top of the tribrach, termed 'centring above the footscrews'.

5.3.4   The horizontal plate(s)

Directly above the tribrach is that part of the instrument known as the *horizontal plate* or *plates*, which may include items 11 to 15 shown in *Figure 5.5*. If there are two plates, they are known as the *upper* and *lower* or *top* and *bottom* plates respectively.

The existence of two-plate construction is indicated by the presence of two horizontal clamps and two slow-motion screws for the control of rotation about the vertical axis. If an instrument has only one clamp and slow-motion screw, then it is of single-plate construction. This type will have an additional control for horizontal movement, either a *circle locking lever* (also termed *repetition clamp*) or a *circle-orienting drive* protected by a hinged cover. These control arrangements affect the methods to be used in setting or reading horizontal angles, shown later.

5.3.4.1   *The horizontal plate bubble*

The upper plate carries a bubble tube (the *plate bubble*) which is used for levelling up the instrument. The traditional procedure, illustrated in *Figure 5.6*, is as follows:

(i) Set up the instrument, lay the bubble tube parallel (in plan) to the line joining any two footscrews.

(ii) Centre the bubble by turning these two footscrews slowly in opposite directions, at the same speed. Note that the bubble will move in the same direction as the user's left thumb.
(iii) Turn the plate through 90° in plan, centre the bubble again, but using the third footscrew only.
(iv) Repeat (ii) and (iii) in the same quadrant in plan until the bubble remains centred.
(v) Turn the plate through 180° – if the bubble stays central, the instrument is levelled up, check in any other position.

Leveiling-up

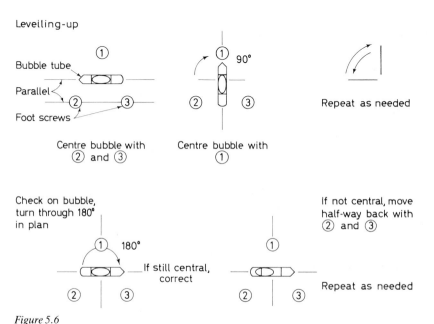

Figure 5.6

If the bubble does *not* stay central, note the position of the bubble with respect to the graduations marked on the tube, see *Figure 5.6*, then:

(i) Move the bubble half-way back to centre using the two footscrews; note its new position against the tube division as the *adjustment position*.
(ii) Turn the plate through 90° in plan, bring the bubble to the adjustment position using the third footscrew.

If the bubble is always brought to the 'adjustment position' on levelling-up, instead of to the centre of its run, then the vertical axis will be properly vertical. A small circular bubble is often attached to the tribrach for rough levelling-up, before the accurate levelling-up with the footscrews.
    In practice, the instrument must be set up and centred over the ground mark as described earlier, then levelled up by the footscrews, then the centring checked again. It may require some repeats before the instrument is both centred and levelled correctly.

5.3.4.2   The optical plummet

Most modern theodolites are fitted with an *optical plummet*, usually within the horizontal plate, as in *Figure 5.7*. This is a small telescope which provides a line of sight down the vertical axis of the instrument, the horizontal view being deflected by a prism. It is an alternative to the plumb-bob for centring the instrument, giving greater accuracy, particularly in windy conditions, but it is only effective if the instrument has been levelled-up so that the vertical axis is truly vertical. Optical plummets may be fitted in tribrachs, or they may be separate, specialized pieces of equipment (see Chapter 7).

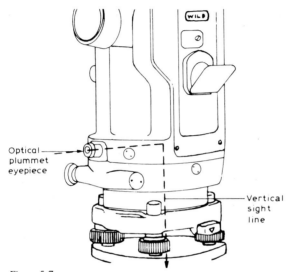

Figure 5.7

If the plumb-bob is detached from the instrument, then the optical plummet and the footscrews may be used both to centre the theodolite and to level it up. The instrument should be set on the tripod with the tribrach roughly horizontal and approximately centred over the ground mark, the optical plummet telescope focused on the ground mark, then, looking through the optical plummet, the footscrews should be adjusted to bring the ground mark central in the field of view. The procedure is then:

(i) Turn the instrument in plan until the bubble tube lies parallel to the line joining the feet of any two legs of the tripod.
(ii) Roughly centre the bubble by extending or shortening one of these two tripod legs.
(iii) Turn the plate through 90°, again roughly centre the bubble by extending or shortening the third leg.
(iv) Level the instrument using the footscrews as described in Section 5.3.4.1.

It will be found that if the tripod legs are firmly positioned in the ground and then when one leg is extended or shortened, the ground mark will often appear to remain stationary when viewed through the optical plummet. Due to this feature, when the above operations have been completed the ground mark will still lie close to the centre of the optical plummet field, and then:

(v) Ease off the clamping bolt and, without rotating, slide the instrument over the tripod head gently to bring the optical plummet cross-hair exactly on to the centre of the ground mark.
(vi) Re-level the instrument as necessary with the footscrews, check the centring, repeat operations (v) and (vi) as needed.

Note that the optical plummet should be checked by turning the plate through 180° and observing whether the cross-hair indicates the same ground point as before. If it does not, then the true plumb point is mid-way between the two indicated positions. As with the bubble tube, this 'adjustment position' should be used until the optical plummet has been adjusted.

5.3.5   The telescope and vertical circle

5.3.5.1   The telescope

The *telescope* was invented in Holland in 1608, the form adopted for surveying instruments being that suggested by Johannes Kepler (1571–1630), known as the *Keplerian* or *astronomical telescope*. The modern *internal focusing telescope* is a tube of fixed length, with an objective lens at the end nearest the object being viewed and an eyepiece lens system at the other. Within the tube is fitted an internal focus lens system, capable of being moved within the tube so as to permit focusing of the distant object. This movable lens is adjusted in position by the movement of a control knob at the side or, more commonly today, by the movement of a sleeve fitted around the telescope barrel.

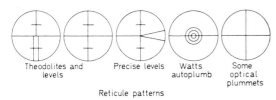

Theodolites and levels   Precise levels   Watts autoplumb   Some optical plummets

Reticule patterns

Figure 5.8

Traditional telescopes provided an inverted image, but it is common practice today to include extra lenses so as to give an erect image, though with some consequent slight loss in light transmission. Telescope lenses are normally coated to reduce reflections and hence improve light transmission. For surveying purposes, the *magnification* and *resolving power* of a telescope are important considerations, magnification typically varying from 10× to 40× depending upon the class of instrument.

The sight line through the telescope is defined by a glass circle with engraved cross lines, inserted between the focusing and eyepiece lens systems. This diaphragm bears a *graticule* or *reticule* of fine engraved lines, commonly called the cross-hairs, since the original diaphragm consisted of a brass ring with spiderwebs stretched across it. Some common arrangements are shown in *Figure 5.8*. The short horizontal lines are *stadia hairs*, used for optical distance measurement, as in Chapters 8, 9, 12, 13 and 14. The intersection of the long horizontal and vertical lines defines the *collimation line* (sight line) of the telescope.

The telescope should be fitted with a pair of external sights, or more commonly today a *finder-collimator* device, used for pointing the telescope initially towards the target to be sighted. This is important due to the narrow field of view of the surveying telescope, often only about 1.5° of arc or a field widtt of about 2.6 m at a distance of 100 m. If the telescope is pointed on target using the sights or collimator and then clamped, when the telescope is focused the target should be visible within the field of view.

5.3.5.2   Focusing

For accurate work when observing a target, both the cross-hairs of the reticule and the distant target must be in sharp focus and achieving this entails two separate operations.

To focus the *eyepiece* on the cross-hairs, point the telescope towards a light background, screw the eyepiece fully in clockwise, then screw it out anticlockwise until the cross-hairs appear sharp and black. To make quite certain, this point should be overrun slightly and then the eyepiece screwed in slightly again to reach the sharp black position. This adjustment depends upon the eyesight of the user and it should be set at the beginning of the day's work. It should not need re-adjustment unless a different observer takes over.

To focus on the *target object*, the focusing knob or sleeve is rotated until the object appears clear and sharp in the plane of the reticule. When this is done correctly, the cross-hairs will appear to be fixed to the target and will not move across the target when the observer's head is moved vertically or laterally. If there is any apparent movement, parallax error exists and if it cannot be eliminated by use of the focusing knob or sleeve then the cross-hairs must be re-focused with the eyepiece as detailed above and the target focused again. The telescope (focusing knob or sleeve) must be re-focused each time the telescope is pointed to a new target, since the position of the focusing lens must be varied as the distance to the target varies.

5.3.5.3   Face left and face right

The transit or trunnion axis carrying the telescope is at right-angles to the telescope collimation line, and it is supported at each end by the standards. (In older instruments the telescope, transit axis and vertical circle could be removed from the standards, but this is not feasible in modern instruments.) Generally one standard is bulkier than the other, since it encases the glass vertical circle mounted on the transit axis. The telescope and vertical circle together rotate in a vertical plane about the transit axis, the movement being controlled by a clamp and slow-motion screw fitted to one of the standards.

The normal observing position when looking through the telescope is to have the vertical circle located at the observer's left-hand side, and this is described as *observing with face left*, simply annotated as FL in the field book. If the telescope is now rotated about the transit axis through an angle of 180°, termed *transiting the telescope* and the plate turned until the eyepiece is at the observer's eye again, then the vertical circle will lie to the observer's right, and this is *observing with face right*, annotated in the field book as FR. Movement in the horizontal plane is controlled by the clamp or clamps and slow-motion screw(s) on the horizontal plate.

In the normal FL position the controls will be found to be in suitable locations for a right-handed observer. It should be noted, however, that angular observations are generally taken on both faces, the results of the two observations being meaned and this process will eliminate or minimize most instrumental errors such as circle graduation error, circle eccentricity, collimation line not at right-angles to the transit axis, or dislevelment of the transit axis, as well as providing a check on the surveyor's readings.

5.3.5.4   The centring thorn

For centring an instrument underneath a mark in the roof of a tunnel, etc., some telescopes are fitted with a *centring thorn*. This is merely a projecting stud on the top of the telescope, such that when the telescope is horizontal the thorn defines the centre of the instrument in plan. The thorn is sometimes connected to a small mirror inside the telescope and turning the thorn adjusts the mirror so as to reflect light on to the reticule and illuminate the cross-hairs in night work or when artificial illumination is being used.

5.3.5.5   The altitude bubble

In addition to the plate bubble, some instruments, are fitted with another bubble tube at the top of the standard supporting the vertical circle. This is the altitude bubble, and it is linked to the vertical circle and to an *altitude*

Coincidence
reader

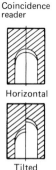

Horizontal

Tilted   *Figure 5.9*

bubble adjusting screw on the standard, similar to a slow-motion screw. When the altitude bubble is central it indicates that the vertical circle is properly zeroed and vertical angles read off the vertical circle will give the correct values. The altitude bubble must be centred by its adjusting screw immediately before taking a vertical angle reading.

The altitude bubble is sometimes read through a *coincidence prism reading system*, instead of being a simple open bubble tube. In coincidence systems, an arrangement of prisms projects an image of half of each end of the bubble to an eyepiece viewer. When the two ends of the bubble appear to be in line, or coincident, as in *Figure 5.9*, the bubble is central and the vertical circle is zeroed. This system allows the bubble position to be set with much greater accuracy than the simple open bubble.

Many modern instruments are fitted with self-zeroing vertical circles (*automatic vertical indexing*) which make use of gravity and avoid the need to centre a bubble before reading a vertical angle. This can speed up vertical angle measurement considerably.

5.3.5.6   Vertical angles

The vertical angle from the instrument centre to a distant point may be measured in two ways. If the angle is measured from the horizontal it is termed an *angle of elevation*, or an *angle of depression*, according to whether the distant point is above or below the horizontal plane through the instrument's centre. In *Figure 5.10* these angles are denoted as + and −. If the vertical angle is measured downwards from the zenith, it is termed a *zenith distance* and denoted here by z. In this connection it must be noted that the direction of the line through the instrument centre to the centre of the earth (downwards) is termed the *nadir*, and the opposite direction vertically upwards from the instrument centre is the *zenith*.

Vertical angles

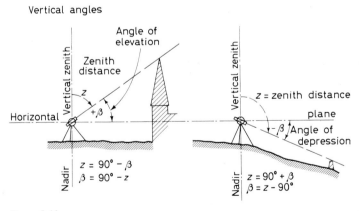

Figure 5.10

5.3.6   Modern theodolite reading systems

The oldest form of theodolite reading system which may still be found in use is the *vernier system*. Such instruments have graduated metal circles which are either completely open or have 'cut-away' windows in casings around the plates and in either case it is possible to see the circle graduations directly without using an eyepiece. In these instruments the readings are made with the aid of vernier scales engraved alongside the index marks, a tricky and time-consuming task. Vernier theodolites are obsolescent now, if not obsolete, and will not be considered further here. The latest systems are electronic, with digital read out facilities, but since these are still under development no serious treatment is possible as yet.

Most of the instruments in use today have *optical reading systems*, where the graduated circles are of glass and are totally enclosed so that circle readings must be made through one or more small telescope-style microscope eyepieces located either alongside the telescope, or on a standard, or on the horizontal plate. These instruments are readily recognized by the small eyepieces, the total enclosure of the circles and the fact that there will be one or more small mirrors to direct light into the instrument to illuminate the circles.

Some instruments have one microscope eyepiece to view the horizontal circle and another to read the vertical circle. In others, a single eyepiece viewer is fitted, either alongside the telescope proper or on one standard. These latter types may show both circles simultaneously or else alternate circle views may be obtained by operation of a changeover knob or switch.

5.3.6.1   Circle microscope system

The circle microscope system is the most basic optical reading system and has a simple fixed hair-line in the microscope, against which the circle graduations are read directly. It is widely used on builder's theodolites.
Example: The Kern KOS construction theodolite with circles graduated to five minutes of arc, thus allowing readings to be estimated to the nearest minute of arc. *Figure 5.11(a)* illustrates a typical set of readings as viewed in the microscope. This instrument has three scales – the lowest being the

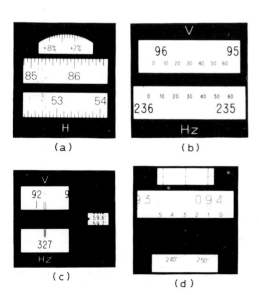

Figure 5.11

horizontal angle, then the vertical and finally a vertical gradient scale showing percentage gradients. (This last is not common on theodolites.) The readings are:

Horizontal 53° 12′; vertical 85° 48′; gradient + 7.35%

5.3.6.2   Optical scale system

Developed from the preceding type, this has a diaphragm fixed in the microscope, with a scale of divisions on it rather than a fixed hair-line.

Example: Figure 5.11(b) shows the scale reading of the Wild T16 general purpose theodolite, with the glass circle divided into single degrees only and the optical scale extending over one degree but divided to single minutes of arc. The number of the degree division is noted, then the number of minutes at the degree division and finally the tens of seconds of arc are estimated (only one degree division can cut the optical scale unless the reading is an exact number of degrees). The readings are:

Horizontal 235° 56′ 20″; vertical 96° 06′ 30″

5.3.6.3   Optical micrometer system

This combines the optical scale principle with lateral movement of the scale under the control of a micrometer setting screw which adjusts the position of a prism in the viewing system. The system is intended to provide greater accuracy than can be obtained with the simpler optical-scale type.

Example: The Wild T1 general purpose theodolite shown in *Figure 5.11(c)* is typical. When the micrometer setting screw is turned, all the numbered graduations appear to move while the central single/double line remains stationary. The micrometer screw is turned until a degree division of the required scale (horizontal (Hz) or vertical (V)) is central between the fixed pair of lines, then the reading is the number of the degree division plus the minutes and seconds (or minutes and decimals of a minute in this example) shown in the third 'window'. It is important to note that a movement of the micrometer screw moves the images of the circles, it does not cause any movement of the circles themselves. The readings are:

Horizontal 327° 59.6′; vertical – the micrometer needs to be re-set.

The presence of a micrometer system is indicated by the extra control knob, the micrometer setting screw, which is neither a clamp nor a slow-motion screw.

5.3.6.4   Coincidence (or double-reading) optical micrometer system

This system is used on precision and universal theodolites where a high accuracy of reading is required, since it provides the mean of two readings taken at diametrically opposite points on the circle, thus minimizing errors of eccentricity. The optical reading eyepiece presents graduations from both sides of the circle and these must be placed equally on either side of a fixed hair-line or, alternatively, brought into coincidence using the micrometer setting screw.

Example: Figure 5.11(d) shows the reading system of the Wild T2 universal theodolite, the upper window showing the graduations brought into coincidence with one another, while the central window shows the degrees and tens of minutes and the lowest window shows minutes and seconds which can be read direct to one second.

The reading is: 94° 12′ 44.4″

5.3.7   Setting on an angle

It is often necessary to point the instrument on a target with a specific reading already set on the horizontal circle. The process is termed *setting on an angle* and the method to be used depends upon the construction of the instrument.

5.3.7.1   *Instruments with upper and lower horizontal plates, each controlled by its own clamp and slow-motion screw*

In these instruments, if only the lower plate clamp is applied then the horizontal circle is clamped to the levelling head and the upper part of the instrument may be rotated about its vertical axis. If the instrument is turned, then the readings on the horizontal circle viewed through the optical reading eyepiece will alter. However, if only the upper plate is

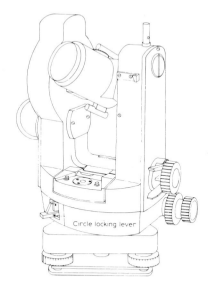

Circle locking lever

Figure 5.12

clamped and the lower plate released, then the horizontal circle is clamped to the upper plate and rotates with the telescope, and in this case the readings will not alter as the instrument is turned. When both plates are clamped it should not be possible to turn the instrument (except by applying excessive force.)

Accordingly, to set on a particular angle, first apply the lower clamp, then turn the instrument until the horizontal circle reading approximates to the required reading and then apply the upper clamp.

If the instrument is a simple circle microscope or optical scale theodolite, use the upper plate slow-motion screw to set the desired reading exactly then unclamp the lower plate. The theodolite may now be turned and the required reading will remain set in the field of view of the optical reading eyepiece; the operation is complete.

If the instrument is an optical micrometer type, then use the micrometer setting screw to set the minutes and seconds and turn the upper plate slow-motion screw until the required degree value is exactly central between the pair of lines (*Figure 5.11(c)*). The required angle is set and the operation is complete when the lower plate is unclamped. It may be noted that some theodolites have a milled rim fitted to the circle, this saving the operator from having to walk around the instrument looking for the desired angle reading.

5.3.7.2 *Instruments with a single horizontal plate clamp, slow-motion screw and a circle locking lever*

The circle locking lever replaces the lower plate clamp and slow-motion screw (*Figure 5.12*) thus the method is basically similar to that in Section 5.3.7.1 above. Various designs of locking lever are used, but in all cases they may be used either to lock the circle to the levelling head or to lock it to the horizontal plate.

To set on a particular angle, lock the circle to the levelling head using the circle locking lever, then release the horizontal plate clamp and turn the instrument until the horizontal circle reading approximates to the required reading, then apply the horizontal plate clamp. Use the horizontal slow-motion screw to set the desired reading exactly, then use the locking lever to lock the circle to the horizontal plate and release the horizontal clamp. The theodolite may now be turned and the required reading will remain set in the field of view of the optical reading eyepiece.

5.3.7.3 *Instruments with micrometer setting screws, circle orienting drive, and no lower plate*

Since the horizontal circle of instruments of this type can only be moved by the circle orienting drive and the circle cannot be swung freely as in the other types, it is not possible to set on an angle before pointing the target.

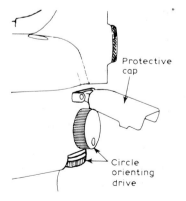

Protective cap

Circle orienting drive

Figure 5.13

Accordingly, before setting on the angle, the target must be bisected, as explained below.

To set on a particular angle, having pointed the target, use the micrometer setting screw to set the required minutes and seconds, then unclamp (or lift the protective cover from) the circle orienting drive and use it to set the required degree value (*Figure 5.13*). The drive should then be re-locked (or the cover replaced) to prevent accidental movement of the circle. Note that in this case, if the horizontal clamp is released and the telescope is swung, the readings will not remain the same, but if the target is re-pointed the angle originally set should re-appear in the reading eyepiece.

5.3.8   Control observations

5.3.8.1   *Horizontal angles*

Various methods of measuring horizontal angles have been used, some being appropriate to older equipment (such as vernier theodolites) or to specialized tasks. For most survey work an observing technique is required which is easy to use, provides some form of check on reading and instrumental errors and will allow an increase in accuracy to be achieved readily when necessary. A suitable method with modern equipment is that known as *simple reversal*, also called *face left and face right on one zero*. For much survey work, measurement on one zero will be sufficient, but an increase in accuracy may be obtained by observing on two or more zeros.

The following additional terms are used in theodolite work and it will be necessary to define them before the observation routines can be fully explained.

Reference Object (RO), back object and *back station* are all terms used to denote the first target the theodolite is pointed at when measuring the horizontal angle subtended at a point or station by two distant target points or stations. *Forward object* and *forward station* are terms used to denote the second target point observed when measuring a horizontal angle.

A *single horizontal angle measurement* is made by pointing on the RO, booking the horizontal circle reading, then turning to point the forward object and again booking the circle reading, all on FL. The difference between the two circle readings is the required horizontal angle and the reading when pointed on the reference object is the *zero* of the measurement. (Remember that theodolite horizontal circles are graduated clockwise from 0° to 360°.) If, as an example, the circle read 45° when pointed on the RO and 65° when pointed on the forward object, then the horizontal angle is 20°, measured from a *zero of 45°*. In practice, the zero could actually be 0°, but this is a matter of choice as will be seen later.

Simple reversal means to measure a single horizontal angle as above, then transit the telescope on to FR, point the forward object again, read and book the circle, then point the RO and read and book the circle. This procedure gives two measures of the angle, the first on FL and the second on FR, the mean of the two values being accepted as the value of the angle if there is no gross error. Note that the pairs of readings (FL and FR) for each target direction provide checks on reading and instrumental errors. A value of an angle obtained in this way, with the graduated horizontal circle kept stationary in one position throughout the operation, is a value obtained by *simple reversal on one zero*.

The accuracy of this first value may be improved by moving the horizontal circle to a new zero setting, repeating the operation to obtain a second value, then taking the mean of the two values. The final value of the angle may be said to have been taken by *simple reversal on two zeros* which, barring gross errors, is more accurate than using only a single zero.

As a general rule then, angles should be measured by simple reversal and for higher accuracy, two or more zeros should be used and the set of results meaned.

Swinging, or *turning*, means rotating the instrument about its vertical axis. Accordingly, swinging (turning) right means rotating the instrument about its vertical axis in a clockwise direction in plan, while swinging (turning) left indicates that the instrument is being rotated in an anti-clockwise direction in plan.

Using the terminology defined, then, a recommended procedure for observing horizontal angles is as follows:

(i) Centre and level-up the instrument over the station, focus the eyepiece on the cross-hairs.

(ii) With the instrument on face left, set on a small angle (the first zero) and sight on to the RO. If the instrument has circle orienting drive, however, sight on to the RO and then set on the small angle.

(iii) Read and record the horizontal angle reading to the RO.

(iv) Unclamp the telescope and horizontal or upper plate, swing right and sight on to the forward object through the open sights or the finder-collimator.

(v) Clamp the telescope and horizontal or upper plate, focus the distant target.

(vi) Bisect the target with the vertical cross-hair, using the horizontal or upper plate slow-motion screws.

(vii) Read and record the horizontal angle reading to the forward object.

(viii) Change face to face right by transiting the telescope.

(ix) Unclamp the horizontal or upper plate and swing left to sight on to the forward object again.

(x) Repeat procedures (v), (vi) and (vii) for a second reading to the forward object on FR.

(xi) Unclamp, swing left to sight on the RO and repeat procedures (v), (vi) and (vii) again to obtain a second reading on the RO on FR.

(xii) Calculate the mean value of the angle.

Four circle readings should have been read and recorded, two on each face, giving two measures of the angle which may be meaned. If desired, the whole process from (ii) above may be repeated with a new zero value set on the circle. If two zeros are to be used, the first could be made just over 00° 00′ 00″ and the second just over 90° 00′ 00″. If four zeros are to be used, they could approximate to 0°, 45°, 90° and 135°, being just over in each case.

Two or more zeros are essential when an accuracy close to the limitations of the equipment is demanded. Thus, when using a one-second theodolite it is necessary to use four or more zeros to ensure that the mean angle is correct to within one second. In addition, it is desirable to use two or more zeros if re-observation would entail a large expenditure of time and effort. In angular observation work generally a disproportionate amount of time is lost in getting to the site, setting up the equipment, etc., while the time spent on observing is minimal. An experienced surveyor would have observed and recorded the four readings above within ten minutes, but it would have taken as long to set up the equipment and still longer if targets for the reference and forward objects had to be established. Travelling time to the site must also be taken into account.

5.3.8.2   Vertical angles

Vertical angle is generally taken to mean the angle of elevation or depression from the horizontal plane to the distant target. Elevation is +ve, depression, below the horizontal is −ve.

A recommended procedure for vertical angle observations is as follows:

(i) Release all clamps, point the telescope at the target with face left, using the open sights or finder-collimator.
(ii) Apply the vertical (telescope) and horizontal circle clamps.
(iii) Focus carefully on the target.
(iv) Bring the intersection of the cross-hairs exactly on to the target, using a horizontal slow-motion screw and the telescope slow-motion screw.
(v) Read and record the vertical circle reading. Note that if an altitude bubble is fitted, this must be centred before the reading is made.

The vertical circle in modern instruments is often graduated from 0° to 360° and when face left the vertical circle reading is actually a zenith distance. Thus with face left, vertical angle = 90° minus the circle reading, but with face right, vertical angle = the circle reading minus 270°.

To eliminate errors, a vertical angle should be observed face left and face right and the two values meaned. Horizontal angles are generally required to be of greater precision than vertical angles and it is therefore common practice to observe all the horizontal angles at a station before commencing the vertical angles, in case any settlement of the tripod occurs. On some occasions, however, such as in poor visibility, it may be expedient to observe the vertical angles first, as this gives the surveyor some familiarity with the target direction.

5.3.8.3   Booking and recording angles

See Chapter 6.

5.3.9   Theodolite and tripod adjustments

From time to time it becomes necessary to make adjustments or carry out maintenance on the equipment. Tripod fittings may become loose, wear may develop in the footscrews, the plate bubble may go out of adjustment, etc. Many of these details are specific to the individual instrument and the maker's handbook should be referred to as to details of what maintenance may be carried out by the surveyor. Certainly a surveyor working in the less populated areas of the world should include the instrument handbook in the survey stores.

Plate bubbles are generally checked and adjusted in the same way as described for the dumpy level in Section 8.7.1, and altitude bubbles may be adjusted by an adaptation of that procedure. If adjustments are required to optical reading systems or the instrument axes, it is recommended that they be carried out by an instrument mechanic specializing in optical theodolites. As with a car, a complex modern theodolite should be serviced occasionally.

5.4   Microcomputer applications

The MICROSURVEY program Lines computes the corrected lengths of survey lines which have been measured with a metric steel tape. According to the user's requirements, corrections may be applied for

(1) slope (based on vertical angle or height difference of ends);
(2) variation from nominal tape length (standardization);
(3) temperature variation from standard; and
(4) local scale factor.

The program can handle up to 25 lines, each consisting of not more than 4 bays (sections at different slopes). Although Lines may be used independently, it may also be used in conjunction with the Traverse program. If the lines of a taped traverse are corrected using Lines, then the program can write the corrected line lengths and the station names to a file which can be read by the MICROSURVEY program Traverse.

When data have been entered in the program the data and results may be stored in a file on disk, and all items of data may be amended, e.g. in case of incorrect data entry or to change a value as a result of re-measurement. Data and results may be displayed and also printed out. Printout may be in summary form, simply listing the lines and their corrected lengths, or in summary and in detail form giving all data, corrections and reduced lengths.

Figure 5.14 shows the summary for a 13-leg traverse, together with part of a later detail page. Note that the user need not calculate standardization, temperature or local scale factors, but merely enters actual and nominal tape length, standard and field temperatures, and the N.G. Eastings of the area.

Leicester Polytechnic

LINE REDUCTION

Survey line reduction printout from CBM micro program LINES

Copy 1   Dated 7th August 1984 (Data held on file l-linesdemo)

Survey name : Leicester Royal Infirmary / No. 4

Client   : R E Paul

Measured by BSc Building Surveying Year 1 on 11th February 1980

Summary of Reduced Line Lengths

Line	0	Stn. 101	to Stn. 402	67.984 metres
	1	Stn. 402	to Stn. 404	61.020
	2	Stn. 404	to Stn. 405	106.763
	3	Stn. 405	to Stn. 406	47.530
	4	Stn. 406	to Stn. 408	25.483
	5	Stn. 408	to Stn. 409	30.610
	6	Stn. 409	to Stn. 410	35.313
	7	Stn. 410	to Stn. 411	26.440
	8	Stn. 411	to Stn. 412	36.420
	9	Stn. 412	to Stn. 413	64.210
	10	Stn. 413	to Stn. 414	55.200
	11	Stn. 414	to Stn. 403	66.736
	12	Stn. 403	to Stn. 206	104.798

(Local scale factor taken as 0.99965 at Nat.Grid Eastings of 460 km)

Continued –

Page 3   Measurement Details

Line 2 from Stn. 404   to Stn. 405

Bay	Slope length	Vert.angle	Ht.diff.	Slope corr.	Hor.length
0	84.448	2 21 20	not taken	-0.071	84.377
1	22.500	1 19 58	not taken	-0.006	22.494

Tape Details

Nominal length   30.000   Actual length   29.983 metres
St. temperature   20.0   Fld. temperature   10.9   deg. C

Line measurement on slope =   106.948
    Reduced to horizontal =   106.870

    Standardisation corr'n =   -0.0588
    Temperature correction =   -0.0109
    Local scale correction =   -0.0374

Final corrected line length =   106.763 metres

Figure 5.14

Chapter 6

Traverse survey

6.1 Introduction

Traversing is probably the most widely used and flexible method of providing control for site surveys, particularly in urban areas where the formation of triangles is difficult and sometimes impossible. A *traverse* is defined as a series of connected straight lines whose lengths and bearings can be determined. The lines are known as legs and the end points of the lines as *stations*.

The legs may be used as chain lines for the supply of detail, or to hold a network of chain lines. Alternatively, detail may be supplied by bearing and distance from the traverse stations. Traverses may commence and close from either end of a base line, from triangulation or trilateration control, or from other existing traverse stations. In the survey of small sites, a traverse often consists of a single closed loop of connected straight lines.

The following sections deal with the techniques to be used in providing control for survey areas by the use of traversing, where the traverses will seldom exceed ten kilometres in length. For longer traverses, or larger areas, the observation, measurement and computation methods are likely to be more sophisticated.

6.1.1 Types and classification of traverse

Traverses are described as being either *closed* or *open*, a *closed traverse* being one whose start and end points (*terminals*) are fixed co-ordinated points. The observations in a closed traverse can be numerically checked and the results mathematically adjusted, but an open traverse cannot be checked or adjusted.

A traverse which commences and closes on the same station is a *closed loop traverse*, sometimes known as a *ring traverse*. A traverse which commences and closes on two different stations, the positions of both of which are already known (co-ordinates for both stations fixed) is a *closed traverse*, sometimes termed a *link traverse*. *Open traverses* should generally be avoided.

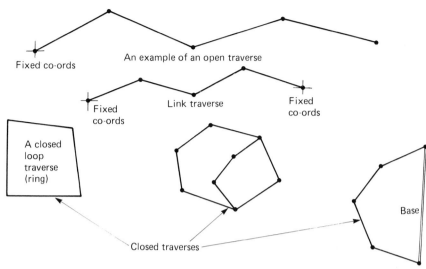

Figure 6.1

Traverses may be classified by the accuracy attained, typically as precise, semi-precise, or rough. *Precise traversing* usually involves the use of a 'single second' theodolite in conjunction with either catenary taping or, more commonly today, electro-magnetic distance measuring equipment. An accuracy considerably greater than 1:10 000 is expected.

Semi-precise describes the majority of traverses. Examples might include the use of a builder's theodolite with a chain, achieving an accuracy of perhaps 1:500, or the use of a 20" theodolite with a steel band perhaps attaining 1:10 000.

Rough traverses indicate an accuracy of less than 1:500, such as the use of a prismatic compass with tape or chain.

At the office stage of the work, traverses may be described as being either *computed* or *plotted*. A *computed* traverse is one in which the field observations and measurements are used to compute rectangular co-ordinates for the stations of the traverse, the traverse then being plotted using the computed co-ordinates. All precise traverses and the majority of semi-precise traverses are computed, since this allows field errors to be located before plotting, gross plotting errors can be avoided, and the accuracy of the traverse can be calculated. A *plotted* traverse is one which is plotted from the field notes, using a scale and protractor. Rough traverses are commonly plotted in this way.

6.1.2   Co-ordinate terminology reminders

The location of a point on a surface may be specified by two measurements, termed *co-ordinates*, but there are several types of co-ordinate systems. The *geographical co-ordinates* of a point on the surface of the earth are the latitude and longitude of the point. The *polar co-ordinates* of a point are its bearing (angle from a specified direction) and its distance from a given pole point.

Mathematical rectangular co-ordinates specify the position of a point by two distances, measured from given x and y axes, the axes being at right-angles to one another. *Surveying rectangular co-ordinates* specify the position of a point by its distance east of a north–south axis (*Easting*), and its distance north of an east–west axis (*Northing*). The north–south axis is the *Reference Meridian*, the east–west axis is the *Reference Latitude*. The method is as used on the Ordnance Survey National Grid, see p. 19.

6.1.3   Bearings terminology

The *bearing* of an observed distant point is the angle between a reference direction and the line from the observer's position to the distant point. A bearing, however, is stated not merely as the angle between the two directions, but as the amount of angular rotation made in turning from the (generally north) reference direction to the direction of the observed point. A *true bearing* of an observed distant point is the angle, measured clockwise, between true north and the line from the observer's position to the distant point.

Similarly, *magnetic* and *grid bearings* are relative to magnetic and grid north respectively. All of these are, on occasion, referred to as 'whole circle bearings', since they may take any value between 00° 00' 00" and 360° 00' 00".

It is most important to note that surveyors and navigators require and use angles clockwise from north (equivalent to the traditional y axis), while mathematicians use angles measured anti-clockwise from the x axis. *Whole circle bearings* may be quoted in the form N 185° E, meaning 'from reference direction north, turn eastwards through an angle of 185°'. The E indicates clockwise rotation, this commencing with an easterly swing from north. *Relative* and *quadrant bearings* are occasionally referred to in surveying. A *relative bearing* is one which is measured relative to some line other than north. An example could be a bearing measured relative to a base line.

A *quadrant bearing* (now obsolescent) is the whole-circle bearing reduced to an angle of less than 90°, so that the 0° to 90° tables of trigonometrical functions can be used in traverse computations. (Today, of course, it is easier to use a pocket calculator with trigonometric functions.) A quadrant bearing is the angle, measured east or west, between the north or south direction and the whole circle bearing direction, for example:

Whole circle bearing   50° = quadrant bearing N 50° E,
                                    100° = quadrant bearing S 80° E,
                                    200° = quadrant bearing S 20° W,
                    and 300° = quadrant bearing N 60° W.

6.1.4   Appropriate uses of traversing

Traverse survey will be a suitable method if:

(1) control is to be provided for a site survey; or
(2) chain survey techniques are inadequate to meet the needs of the task; or

(3) the accuracy demanded is greater than can be achieved by chain survey; or

(4) the introduction of traversing into a chain survey will reduce the costs and/or the duration of the job, by reducing the number of tie and check lines which would normally be needed.

6.2   Traverse reconnaissance and layout

On arrival on site, the surveyor should first carry out a thorough examination of the area to determine the best positions for the traverse stations and hence the traverse legs. Time spent on such a 'recce' is seldom wasted and generally leads to a more satisfying end product. A number of factors must be considered in making the choice of station and leg positions, identified in the following sections.

6.2.1   Preliminary considerations

These may include:

The need to tie the traverse to higher order control in order to meet the accuracy demanded or help locate any errors.

The equipment available and its characteristics, e.g. if taping the legs, then they should preferably run over the more level ground, avoiding obstacles.

The climatic conditions, e.g. mist and haze, may limit leg lengths.

The quality and quantity of the labour to be used.

The time available for the task.

The method of supplying detail. If offsets from the legs are used, the legs must run close to the detail. If bearing and distance is used, there must be a clear view of the area to be surveyed from any station.

6.2.2   Traverse layout

A *closed loop traverse* should preferably be regular in shape, e.g. a perfect square, hexagon, etc. This is seldom practicable, but large length differences and acute direction changes in consecutive legs should be avoided, in order to simplify the distribution of the small errors which can be expected in the observations and measurements of the traverse.

The *legs* should be as long and as few as possible, bearing in mind the equipment to be used and the climatic conditions. This will usually minimize the observational, computational and plotting time. Legs of less than 20 m in length should be avoided, due to the effects of centring errors on such legs and the distribution of errors through these legs. Legs should be located along the more level or even ground, avoiding obstacles, steep slopes, rough ground and marsh if distances are to be taped on the surface.

Stations should, if possible, be placed on permanent ground features, e.g. the inside corner of a manhole cover frame, since this simplifies their location. Artificial station markers might include a wooden peg with a nail driven in the top to mark the exact station point, or a steel pin placed in concrete, in either case with possibly a wooden marker peg placed alongside bearing the station identification in waterproof crayon or paint.

Station markers should be protected from damage, either by 'fencing-off' or by suitable location selection. The type of marker depends upon the permanence required. If station markers are likely to be disturbed, they should be *referenced* to other pegs or permanent objects, and the referencing noted in the field book. To *reference a point*, note its distance from several permanent marks, thus the point may be relocated later by measurement from the permanent marks.

6.2.3   Tying to other control

If a closed loop traverse is inadequate for a survey task, then a network of secondary traverses may be a possible alternative. These 'secondary' traverses should all lie within the 'primary' loop traverse. A *secondary traverse* should be as straight as possible between its opening and closing stations (terminals) and all its legs should lie within the circle which has the line joining the traverse terminals for a diameter, as in *Figure 6.2*.

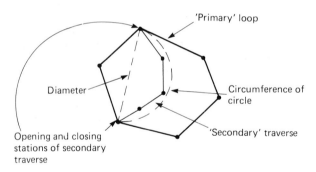

Figure 6.2

Adjacent traverses in a network which run close to one another should, if at all possible, be linked at their mid points, as in *Figure 6.3*. This is to ensure that the detail from both traverses is compatible, e.g. the north face of a rectangular building could be supplied from one traverse and the south face from the other traverse.

Where a traverse is linked to other control, either a primary loop as mentioned above or higher order traverse or triangulation, the terminal points of the traverse would obviously be known co-ordinated points on the existing control. The co-ordinates of the terminals being known, then the traverse angles and the computed traverse co-ordinates may be checked and adjusted. On occasion, when tying to existing control, there may be problems such as an existing point which can be measured to, but which the theodolite cannot be stationed over. This and similar problems are considered in Section 6.6.

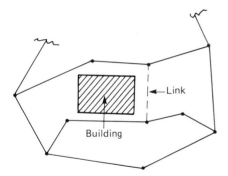

Figure 6.3

6.3   Angle and distance measurement

In *semi-precise traversing*, three distinct standards of accuracy may be required, according to the nature of the task. Where a traverse is run solely for the purpose of supplying *detail for plan production*, a closing accuracy of 1:500 is normal, since this standard is acceptable for the plotting of detail and is generally suitable when area measurements from plan are necessary. Where a *network of traverses* is required, or a closed loop traverse holds a network of chain lines, it is usual to aim for a greater accuracy, possibly 1:5000.

Traverses from which detail is to be set out in preparing for building or engineering work generally require still higher standards of accuracy. British Standard BS 5964: 1980, *Methods for Setting Out and Measurement of Buildings: Permissible Measuring Deviations*, specifies that such traverses should close to within $\pm 2\sqrt{L}$ mm, where L is the length of the traverse in metres, and this is normally better than 1:5000. The following sections outline the procedures recommended if these standards are to be achieved.

6.3.1   Setting out traverses

Each *leg* should be measured at least twice, using a steel band, with readings to the nearest millimetre and corrections applied for standard, temperature, slope and tension. If tied to Ordnance Survey control, the measured distance should also be corrected by the O.S National Grid local scale factor.

The two *field measurements* of any line length should agree within 2 mm/30 m of tape or part tape length. It may not be necessary to standardize the band if it is fairly new and undamaged. Long legs over undulating ground should be measured in catenary and corrected for sag. Alternatively, legs may be measured by electromagnetic distance measuring equipment or with self-reducing tacheometers and horizontal staves, see Chapters 12 to 14.

Horizontal angles should be observed with a theodolite, with readings booked to at least the nearest 20″, on face left and face right (the simple reversal technique) and the reduced readings should agree within one minute of arc. Where traverses are of considerable length (say 1500 m or more) the readings should be taken on two or more zeros. On long traverses with few legs this is essential to maintain accuracy, and on long traverses with many legs this may eliminate the need for later re-observation, which could entail considerable expenditure in time and money.

The *angle of slope* (vertical angle) between the traverse stations and/or changes of slope are often measured by the theodolite, but the Abney level may be used or a level and staff, in all cases for the purpose of reducing the slope leg length to the horizontal. The *vertical angles* are usually observed on completion of the horizontal angles. They are measured in one direction only, but on both left and right faces. Alternatively, they may be read on one face only, but in both directions, i.e. both up and down the slope. The vertical angles are often read and booked to the nearest 20″ of arc, but on gentle sloping sites readings to the nearest minute are adequate.

The two reduced vertical angles for each leg should agree as follows:

For slopes less than ± 5°, within 5′ of arc, slopes 5° to 20°, within 1′ of arc, and slopes over 20°, within 20″ of arc.

6.3.2   Traverses controlling a network

The *linear measurement procedures* required are similar to those for setting out traverses, but some economy of effort is acceptable. Each *leg* should be measured twice, with a band which has been standardized (unless fairly new and undamaged). Readings to the nearest millimetre, corrections for standard, temperature and slope. Some tension should be applied, but it is not usually necessary to use a spring balance. If tied to O.S. control, local scale factor correction is required.

Taping in catenary may be avoided, except in short lengths up to 15 m, when the sag correction may be ignored because it is very small. The two measures should agree to the same limits as before. Again, EDM or self-reducing horizontal bar tacheometer measurements are suitable. Angular measurements should follow the same procedures as for setting out traverses, but substitute 30″ for 20″, and angles of slope of less than 10 minutes of arc may be ignored.

6.3.3   Single detail traverse

The linear measurement may be carried out as in chain survey, but it is preferable to use a steel tape instead of a chain and read and book distances to the nearest 0.05 m.

Horizontal angles should be read and booked to the nearest minute of arc on both faces, the reduced readings should agree within two minutes of arc.

Slope correction may be made as in chain survey.

6.4   Traverse data recording

Traditional traverse booking may be carried out using loose leaf booking sheets or a field book, in either case A4 size is recommended. The methods used for booking vary, but most surveyors tend to use one of the styles illustrated below. As with all booking, of course, clarity and accuracy are essential.

6.4.1 Detail traverses

In *detail traversing*, it is common practice to record the horizontal leg lengths, together with any necessary offsets, in the chain survey booking. Accordingly, a single booking sheet can be used, recording all the angular observations in the traverse but with no linear measurements shown, as in *Figure 6.4*.

The form is fairly self-explanatory, but a few points may be noted:

RO = *reference object*. This is the object on which the theodolite is pointed first when starting to observe a set of directions. The horizontal circle readings are recorded in the order in which they are observed, working downwards from the top of the page. Three different 'zero setting' techniques are illustrated in the figure, although in practice only one, the surveyor's own preference, would be adhered to.

In *example 1*, at station A, pointing the RO on face left, the circle has been set at 00° 00'.

In *example 2*, at station B, a small angle of 03° 10' has been set on the instrument.

In *example 3*, at station C, no particular angle was set on, the theodolite was simply pointed at B and the reading noted, and in this case it happened to be 50° 14'. This method was also followed at station D, pointing on C, but in this case the reading is rather large and is likely to make the subsequent arithmetic subtraction work laborious.

Note that at each instrument station the FR readings should differ from the FL readings by approximately 180°. The difference would be exactly 180° if the observations were free from all error or if the errors compensated one another.

The 'Reduced to RO' column gives the results of subtracting the RO reading from the forward reading for each consecutive pair of readings. Thus at A, FL is 80° 27'−00° 00' = 80° 27', FR is 260° 28'−180° = 80° 28'. At B, FL is 99° 01'−03° 10' = 95° 51', and so on.

The traverse is a closed loop, and the readings have been booked moving round the stations in an anti-clockwise direction in plan, thus the angles reduced to RO are the internal angles of the figure. Since the internal angles of a polygon sum to $(2n - 4)$ right angles, where n = the number of sides, the angles should sum to 360°. This may be checked on the booking sheet and the meaned values adjusted equally before the traverse computation, if desired. (This check should be carried out before leaving the site, to detect any gross error.) The traverse could have been carried out in a clockwise direction, then the reduced angles would be the external angles of the figure, which should sum to $(2n + 4)$ right angles.

Inst. stn	Face	Station observed	Horizontal reading	Reduced to RO	Mean	Corrected angle
A	L	RO. Sh. D	00° 00'			
		Sh. B	80 27	80° 27'	80° 27½'	80° 27'
	R	Sh. B	260 28	80 28		
		RO. Sh D	180 00			
B	L	RO. Sh. A	03 10			
		Sh. C	99 01	95 51	95 51½	95 51
	R	Sh. C	279 01	95 52		
		RO. Sh. A	183 09			
C	L	RO. Sh. B	50 14			
		Sh. D	136 09	85 55	85 55	85 54
	R	Sh. D	316 09	85 55		
		RO. Sh. B	230 14			
D	L	RO. Sh. C	327 29			
		Sh. A	65 19	97 50	97 49	97 48
	R	Sh. A	245 18	97 48		
		RO. Sh. C	147 30			
				Sum.	360° 03'	360° 00'
				Misclosure	03'	✓
			Correction= -¾' per station			
		(Nte each corrected angle has been rounded up or down to the nearest minute)				

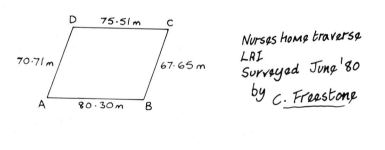

D   75·51m   C

70·71m   67·65 m

A   80·30m   B

Nurses Home traverse
LRI
Surveyed June '80
by C. Freestone

Figure 6.4

| Traverse points | | Horizontal angles | |
Inst.	Target	Face Left ° ′ ″	Face Right ° ′ ″
B	A	03 10	183 09
	C	99 01	279 01
Angles = Mean =		95 51 95 51 30	95 52

Figure 6.5

Figure 6.5 shows an alternative form of presentation for one 'round' of horizontal angles. The observations at Station B in *Figure 6.4* are illustrated.

6.4.2 Other theodolite traverses

When a traverse is to be tied to other traverses, or if the individual station details are more complex than with a simple detail traverse (e.g. observations on more than one zero, vertical angles by theodolite and distances corrected for temperature, standardization, etc.), then it is often better to have a separate sheet or page for each traverse station's booking. An example is shown in *Figure 6.6*. As an alternative, one book could be used for angular measurement recording and another for linear measurement details.

Again the form illustrated is fairly self-explanatory, but the following points should be noted:

The *horizontal angle readings* are at a station identified as 209, on a traverse known as LRI − TWO. This station is also the opening station of traverse LRI − THREE, which was observed on completion of traverse LRI − TWO. The observations have been taken to station 303 in order to avoid the need to re-occupy station 209 at a later date, i.e. when traverse THREE is commenced. Note that for the first zero, the observer has actually used 00° 00′ 00″, but the zero for the second round is a completely different, arbitrary setting of 71° 12′ 20″. The 'Mean' column shows the means of the four readings for each direction.

The *reduced vertical angle* is the angle of slope of the leg (or part) concerned. The calculation depends on the mode of graduation of the vertical circle of the theodolite in use. In this instrument, the vertical circle is designed to read 90° 00′ 00″ when the telescope is horizontal on face left, and 270° 00′ 00″ on face right. Thus the reduced vertical angle to station 210 is 90° 00′ 20″ less 90°, giving a slope of −00° 00′ 20″ on FL, and on FR 270° 00′ 10″ less 270°, giving +00° 00′ 10″, with a mean value of −00° 00′ 05″.

Traverse	LRI – Two		Instrument	Wild T16
Date	20 Nov '79	Weather Fair	Observer	Greasby
Horizontal angles at station		209	Booker	Heminsley

Station observed	Face / Zero	Horizontal reading	Reduced to RO	Mean
RO. Stn 208	L/1	00° 00′ 00″	° ′ ″	179° 25′ 38″
210		179 25 40	179 25 40	247 31 55
303		247 31 50	247 31 50	
303	R/1	67 32 00	247 31 40	
210		359 26 00	179 25 40	
RO. Stn 208		180 00 20		
RO. Stn 208	L/2	71 12 20		
210		250 37 40	179 25 20	
303		318 44 10	247 31 50	
303	R/2	138 44 20	247 32 20	
210		70 37 50	179 25 50	
RO. Stn 208		251 12 00		

Vertical angles at station **209**

Stn. obs'd	Face	Vertical rdg.	Reduced angle	Mean	Observer Greasby
210	L	90° 00′ 20″	– 00° 00′ 20″	– 00° 00′ 05″	Booker
	R	270 00 10	+ 00 00 10		Heminsley
c/s I	L	90 15 20	00 15 20	00 15 05	Date
	R	269 45 10	00 14 50		20 Nov '79

Surface taped distances – Tape No. **5**   Length of tape **30 m**

Line	No. of full tape lengths laid	Final tape rdg.	Total measured length	Temp °C	Remarks, sketch or change of slope diagram
209-208	✓✓✓✓	1·344	121·344	3	
209-210	✓✓✓✓	1·239	121·239	3	
209-303	✓✓✓	21·800	111·800	3	

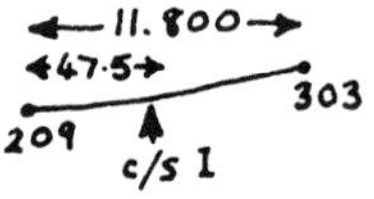

Booker Heminsley
Date 23 Jan '80

Figure 6.6

Note that when the line of sight is nearly horizontal, it often happens that one reduced angle is positive and the other negative. For clarity, the signs have been entered here, but they are not essential – see the second result of 00° 15′ 05″.

The measured length of the leg from 209 to 208 is quoted as 121.344 m, i.e. 4 × 30 m lengths + the final tape length of 1.344 m, all measured at 3°C. Note that there is a change of slope between stations 209 and 303. Distance measurements to the nearest 0.25 m are usually adequate for the length from the station to the change of slope point, 47.5 m in the example. Remember, however, that the total length should be recorded to the nearest millimetre. Note also that the leg 209 to 303 is still booked to the nearest millimetre. Had 21.8 been booked, the individual doing the computation would be uncertain as to whether the distance had been measured to three decimal places.

6.5   Traverse computation

The minimum equipment for traverse computation today should be a pocket calculator with trigonometric functions, preferably including polar–rectangular and rectangular–polar conversion, and a square-root key. If several traverses are to be computed, a programmable calculator is to be preferred, ideally with program cards and a printer, or better still a microcomputer with suitable software and peripherals.

Before the actual calculation of co-ordinates from the observed angles and measured distances, the steps outlined in the next three sections may require to be carried out.

6.5.1   Reduction of measured lengths

The *field measured leg lengths* must be reduced to the *corrected horizontal equivolents*, by applying corrections for slope, temperature, standardization, etc., as covered earlier. *Figure 6.7* illustrates part of a typical 'Reduction of the measured length' form which could be used with a pocket calculator. The following points should be noted:

The correction factors for standardization and temperature have been obtained from the formulae in Sections 5.2.2.1 and 5.2.2.2 respectively.

The projection correction factor may be calculated to five decimal places using the formula

$$0.999\,601\,5 + [(E - 400)/9027]^2$$

in which E = National Grid Eastings in kilometres. A more precise formula is given in the publication Ordnance Survey 1950, *Projection Tables for the Transverse Mercator Projection of Great Britain*, HMSO, London.

Taking the first leg (505–506) as an example, it will be seen that the leg has been measured twice (41.800 and 41.798). These measurements have been meaned, then the result multiplied, in turn, by the correction factors for *standardization* (giving 41.7942), for *temperature* (giving 41.7906), then by the *O.S. scale factor* (giving 41.7760), and finally by the cosine of the vertical angle to give the *reduced horizontal distance* (here 41.753 m).

REDUCTION OF THE MEASURED LENGTHS

~~Base~~/ Traverse   *LRI - FIVE*

Calculated by _____ *R Jones* _____ Date _3 Mar '80_

Correction factors for:

STANDARD		TEMPERATURE		PROJECTION		
Tape No.	Factor	°C	Factor	N.G Eastings	Factor	
⑤	0.99990	11	0.99990	~~450 000~~	~~0.99963~~	
⑦	0.99987	14	0.99993	460 000	0.99965	
				~~470 000~~	~~0.99966~~	

Leg	Measured	Corrected for:			~~Cos~~. V A for Slope	Reduced Horizontal Distance
		Standard	Temperature	Projection		
505 -506	41·800 41·798	⑦ 41·7942 ⑤	11°C 41·7906 14°C	41·7760	01°53'28"	41·753
506 -507	45·850 45·848	⑦ 45·8437 ⑤	11°C 45·8398 14°C	45·8238	00°58'52"	45·817
507 -602	36·810 36·815	⑦ 36·8083 ⑤	11°C 36·8051 14°C	36·7923	00°24'58"	36·791

Figure 6.7

6.5.2   The traverse abstract

The *horizontal angles* at each traverse station, measured clockwise from the back station to the forward station (i.e. the angle included between the pair of legs meeting at the station, hence known as 'included angles') are readily obtained from the field bookings, but if there are any complications it may be better to prepare an *abstract of the angles*. The complications referred to may result from the use of more than one booking sheet for a station, the employment of two surveyors on the same task, or difficulties in tying the traverse to other traverses.

A part of such an abstract is shown in *Figure 6.8*. The entries under '1 st zero' are the horizontal angular readings reduced to RO, the 2 nd and 3 rd zero columns indicating seconds only. In this case, the reduced angles on 2 nd and 3 rd zero differ from the 1 st zero values only in the number of seconds in each, but if there were a difference in the numbers of minutes also, then the minutes values would also be shown.

On the traverse booking sheet, *Figure 6.6*, the booker meaned both zeros and this meaning of each zero is a preferable alternative if an abstract is to be used. Note that additional readings may be taken and entered when the horizontal angle readings and/or the measured distances do not conform to the job specification.

6.5.3   Allocation of opening co-ordinates and bearing

Before starting to compute the traverse, the co-ordinates of the first or start station of the traverse (the 'opening co-ordinates') and the bearing of the first leg ('opening bearing' or 'initial bearing') must be known.

TRAVERSE ABSTRACT   Leicester Royal Infirmary   Page 1 of 2.

Traverse Four   Date 11/2/80   Abstracted by Louise Smith

Inst Stn.	Stn. Obs'd.	1st zero ° ' "	2nd	3rd	4th	Mean ° ' "	Measured Dist.	Tape No.	Temp °C	Vertical Angle
101	Cath.	00 00 00	00			00 00 00				
	402	165 06 55	50			165 06 52	68.063	2	16	01° 11' 15"
	476	259 24 60	50			259 24 50	68.065	5	14	01° 11' 20"
402	101	00 00 00	00			00 00 00				
	404	107 32 20	3152			107 32 06	61.103 / 61.090	2 / 2	15 / 10	01 18 45
404	402	00 00 00	00			00 00 00				
	405	281 58 55	30			281 58 42	106.960 / 106.945 / 106.973 / 106.917	2 / 2 / 5 / 2	10 / 6 / 4 / 4	01 19 58
504	404	00 00 00	00			00 00 00				
	404	178 23 45	226 23 48	226 23 48		175 23 23	47.650	2	13	01 48 08
	413	163 13	181	142		163 13 45	64.280 / 64.268	5 / 2	14 / 10	00 01 58

Tape constants etc.

2 = 0.999008
5 = 0.99990

Figure 6.8

If the traverse commences on a previously co-ordinated point, such as an O.S. or other control point, then the opening co-ordinates can be obtained and the opening bearing may be deduced. (Section 6.6 deals with a variety of problems which may arise when tying a traverse to existing control.)

If the traverse also closes on a previously co-ordinated point, then it will be possible to check and adjust both the angles of the traverse and the calculation of the co-ordinates. Where the traverse is independent of existing control, but is to be plotted by co-ordinates, then the surveyor must decide what values to use for the opening co-ordinates, and what opening bearing to use.

6.5.3.1   Selection of opening co-ordinates

Methods commonly used to select *independent opening co-ordinates* are as follows:

(i) Take the first station as simply 0.000 m East, 0.000 m North. The disadvantage of this method is that it is very likely that some stations will have negative co-ordinates, and matters are simpler for the beginner with all positive co-ordinates.

(ii) Select opening co-ordinates such that all stations in the traverse will have positive co-ordinate values. In *Figure 6.9*, Station A has been given opening co-ordinates of 100.000 m E, 100.000 m N, thus all the stations lie in the north-east quadrant and all co-ordinate values are positive. This removes the arithmetic problems of method (i).

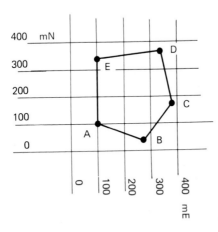

Figure 6.9

(iii) Scale the co-ordinate values for the start station from an existing gridded map, e.g. an O.S. map, and use the rounded scaled figures. For example, scaled O.S. values for the point might be 410895.2 m E, 286116.6 m N, then the surveyor might decide to use 895.000 m E and 6117.000 m N for his opening co-ordinates. This has the advantage of giving the client some compatibility between the O.S. map and the new plan or map.

6.5.3.2   Selection of the opening bearing

The *opening co-ordinates* fix one point of the survey on the grid, while the *opening bearing* decides the orientation of the survey. Opening bearings often used for independent traverses are as follows:

(i) Grid north (or south, east of west), for the particular grid in use.

(ii) The compass bearing of the opening or closing leg of the traverse, possibly corrected for magnetic variation. (Variation, also known as declination, is the horizontal angle between true north and magnetic north at a place, and it may be east or west of true north. Variation differs from place to place, and is subject to small annual and diurnal changes, e.g. in 1983 in eastern England it is about 5.5° W, changing by 0.5° E every three years, while in the south-west it is about 8.5° W and changing by about 0.5° E every eight years. Some local and national map sheets give this information.)

(iii) The approximate bearing of the first leg, as plotted roughly on an existing map and measured by protractor, or by scaling co-ordinates and calculating an approximate grid bearing. Again, this gives some compatibility between the existing map and the new plan or map.

6.5.4   Calculation and adjustment of traverse bearings

The *whole circle bearings* of all the legs of the traverse are required before the co-ordinates of the stations can be computed. The process of calculating the bearings from the observed angular data is often termed 'working through the bearings'.

6.5.4.1   Forward and back bearings

At any traverse station, the next station ahead is termed the *forward station*, and the preceding station is termed the *back station*. In a closed loop traverse, every station has a back and a forward station, of course, but in a link traverse the back station from the first station will generally be an RO, reference object, from which the opening bearing will be determined. Similarly, in a link traverse, the forward station from the last traverse station will be a *closing RO*.

At any given station, the whole circle bearing of the leg to the back station is termed the *back bearing* at the given station and the whole circle bearing of the leg to the forward station is the *forward bearing* at the given station.

Figure 6.10 shows station C, and the back bearing at C is the whole circle bearing of the line from C to B (the back station), while the forward bearing is the whole circle bearing of the line from C to D (the forward station). It will be evident that:

The back bearing at C (bearing of line C–B) + the angle observed at station C (angle BCD) = the forward bearing at C (bearing of line C–D).

If the bearing C − B was deduced as in Section 6.5.3, then addition of the observed clockwise angle at C gives the forward bearing C−D.

Inst. stn	Face	Station observed	Horizontal reading	Reduced to RO	Mean	Corrected angle
A	L	RO. Sm. D	00° 00'			
		Sm. B	80 27	80° 27'	80° 27½'	80° 27'
	R	Sm. B	260 28	80 28		
		RO. Sm. D	180 00			
B	L	RO. Sm. A	03 10			
		Sm. C	99 01	95 51	95 51½	95 51
	R	Sm. C	279 01	95 52		
		RO. Sm. A	183 09			
C	L	RO. Sm. B	50 14			
		Sm. D	136 09	85 55	85 55	85 54
	R	Sm. D	316 09	85 55		
		RO. Sm. B	230 14			
D	L	RO. Sm. C	327 29			
		Sm. A	65 19	97 50	97 49	97 48
	R	Sm. A	245 18	97 48		
		RO. Sm. C	147 30			
				Sum.	360° 03'	360° 00'
				Misclosure	03'	✓
		Correction = -¾' per station				
		(Note each corrected angle has been rounded up or down to the nearest minute)				

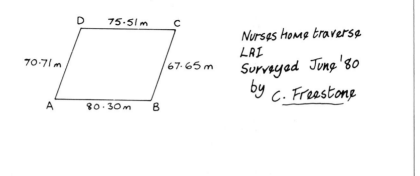

D 75.51m C

70.71m 67.65m

A 80.30m B

Nurses Home traverse
LRI
Surveyed June '80
by C. Freestone

Figure 6.4

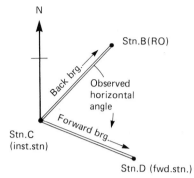

Figure 6.10

Considering station D, the back bearing D−C is equal to the forward bearing C−D plus or minus 180°, then the process above may be repeated to obtain the forward bearing D−E, and so on.

6.5.4.2   Closed loop traverse

Figure 6.11 illustrates the closed loop traverse booked in *Figure 6.4*, repeated opposite. The surveyor has decided to take the closing bearing of the traverse (leg A−D) as grid north, then the bearings of the traverse are deduced as shown. Note that in the numeric example the ± 180° has been

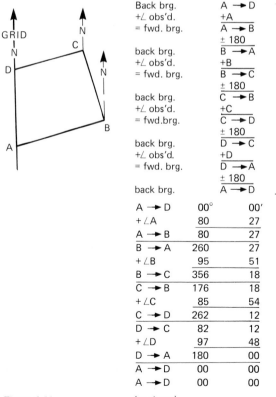

Back brg.	A → D
+∠ obs'd.	+A
= fwd. brg.	A → B
	± 180
back brg.	B → A
+∠ obs'd.	+B
= fwd. brg.	B → C
	± 180
back brg.	C → B
+∠ obs'd.	+C
= fwd.brg.	C → D
	± 180
back brg.	D → C
+∠ obs'd.	+D
= fwd. brg.	D → A
	± 180
back brg.	A → D

Note the repeating patterns:

back brg. + ∠ obs'd = forward bearing

and the grouping of the station letters

AAA, BBB, CCC, etc.

If the observed angles entered are the corrected ones, then the first line should equal the last line as is shown below

A → D	00°	00'
+ ∠A	80	27
A → B	80	27
B → A	260	27
+ ∠B	95	51
B → C	356	18
C → B	176	18
+ ∠C	85	54
C → D	262	12
D → C	82	12
+ ∠D	97	48
D → A	180	00
A → D	00	00
A → D	00	00

Figure 6.11   (as above)

done, but the figures have not been written in. The user may prefer to show the figures.

The rule as to whether to add or subtract the 180° is:

if the forward bearing is less than 180, add 180,
if the forward bearing is greater than 180, deduct 180,
to obtain the back bearing at the next station.

On occasion the summation will give a value greater than 360°, in this case 360° must be subtracted to get the correct value.

As a general rule, a small sketch of the traverse should be drawn, to help check that the values determined for the bearings are reasonable.

6.5.4.3   Closed traverse tied to other control

Figure 6.12 illustrates a traverse entitled 'Gateway Traverse', commencing on station 01 at one end of a base line and closing on to station 12 in another traverse called 'Base Traverse'. In this case, the calculations for

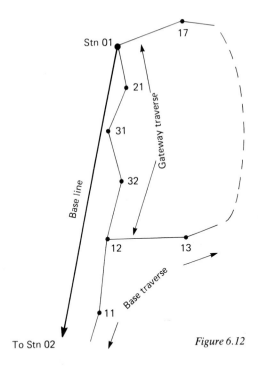

Figure 6.12

the Base Line and for the Base Traverse will have to be carried out before the Gateway Traverse can be computed, since the Base Line and Base Traverse will fix the co-ordinate values for the start and end stations (01 and 12) of Gateway Traverse.

(i) Opening bearings: Given that the co-ordinates of stations 01 and 02 have already been calculated, then the bearing of the line 01–02 can be calculated, in this case it is 180° 00′ 00″, and this, together with the angle observations at station 01, will give a bearing for the first line of the traverse, 01–21, the 'opening' or 'initial' bearing.

Similarly, however, the co-ordinates of stations 01 and 17 will allow the calculation of the bearing of the line 01–17, found to be 63° 13′ 08″ here, and from this another value for the opening bearing of 01–21 may be deduced. Note that the method of calculating bearings from co-ordinates is covered in Section 6.6, but this need not be considered further at this stage.)

It might seem that only EITHER the bearing 01–02 OR the bearing 01–17 need be used to find the bearing of line 01–21, but the calculation of two values for the bearing, from different data, provides a check on gross error. Generally, where two or more opening bearings are available, then either the mean value will be used or the value which is considered to be the most reliable. Note that bearings to distant points are considered usually to be more reliable than bearings to nearby points.

TRAVERSE ABSTRACT *City Campus*
Traverse *Gateway*   Date *16/6/83*   Abstracted by *R Drusary*

Inst Stn.	Stn. Obs'd.	1st zero ° ′ ″	2nd	3rd	4th	Mean ° ′ ″	Measured Dist.	Tape No.	Temp °C	Vertical Angle
01	02	00 00 00	00	00	00	00 00 00				
	17	243 13 00	00	10	00	243 13 03				
	21	347 00 00	00	10	10	347 00 05				
21	01	00 00 00	00	00	00	00 00 00				
	31	195 13 00	00	10	13/40	195 12 55				
31	21	00 00 00	00	00	00	00 00 00				
	32	190 46 60	50	50	30	190 46 48				
32	31	00 00 00	00	00	00	00 00 00				
	12	154 14 50	40	40	50	164 14 45				
12	32	00 00 00	00	00	00	00 00 00				
	13	106 47 20	00	20	30	106 47 18				
	11	197 56 10	10	40	40	197 56 25				

Figure 6.13

(ii) Closing bearings: When an opening bearing for the traverse has been decided, the bearings of the traverse will be worked through, and, to provide a check on these calculated bearings, a 'closing bearing' must be determined in a very similar way to the above. Here, using the co-ordinates of stations 12, 11 and 13, closing bearings were calculated for lines 12–11 and 12–13, and these were 185° 12′ 40″ and 94° 03′ 32″ respectively. Working through the traverse bearings will give bearings for these lines also and they may be compared with the co-ordinate-based bearings above. The comparison will determine the angular misclosure of the traverse.

(iii) Working through the Gateway Traverse: An abstract of the observed horizontal angles for the Gateway Traverse is shown in *Figure 6.13*. Working through the bearings produces the following result:

Opening bearing	01→02	180° 00′ 00″	... from co-ordinates
	+∠01	347 00 05	... from traverse abstract
	01→21	167 00 05	... 180 + 347 = 527 527 − 360 = 167
Opening bearing	01→17	63 13 08	... from calculated bearings
	+∠01	103 47 02	... from traverse abstract 347° 00′ 05″−243° 13′ 03″
(Check only)	01→21	167 00 10	... Check reading only, 01→02 is longest leg
Back bearing	21→01	347 00 05	
	+∠21	195 12 58	
	21→31	182 13 03	... 347 + 195 = 542, 542 − 360 = 182
	31→21	02 13 03	
	+∠31	190 46 48	
	31→32	192 59 51	
	32→31	12 59 51	
	+∠32	154 14 45	
	32→12	167 14 36	
	12→32	347 14 36	
	+∠12	106 47 18	... the clockwise angle at 12 between 32 and 13
	12→13	94 01 54	... again −360°
Closing bearing	12→13	94 03 32	... from co-ordinates
	Misclosure	−01′ 38″	
	12→32	347 14 36	... repeated from above
	+∠12	197 56 25	... the clockwise angle at 12 between 32 and 11
	12→11	185 11 01	... again −360°
Closing bearing	12→11	185 12 40	... from co-ordinates
	Misclosure	−01′ 39″	

Note that the comparison of the two bearings for line 12–13 shows a misclosure of −01′ 38″, while comparison of the two bearings for line 12–11 give a misclosure of −01′ 39″. Where two or more misclosures are calculated, the surveyor must normally decide whether to accept one of them as the closing error or take their mean. In this case, the two are practically identical.

For an accuracy of 1:2000, a closing error of $\sqrt{n}'$ is acceptable, where n = number of traverse stations, and for 1:5000 then $30\sqrt{n}''$ may be satisfactory. If the traverse is short these values may be exceeded provided the co-ordinate misclosure is acceptable, i.e. traverse co-ordinates closing to within 1 part in 2000 or 1 in 5000 respectively (see Section 6.5.7).

In the setting out of buildings, BS 5964:1980 suggests an angular misclosure limit of $\pm 0.135/\sqrt{L}°$ where L is the traverse length in metres. The Gateway misclosures of 98 and 99″ are approximately equivalent to $44\sqrt{5}''$, which is a little in excess of the technical instructions issued to the surveyor. However, the traverse was short, roughly 200 m and only four legs, and hence it was decided that the result was acceptable.

The two misclosures here are along closing ROs of approximately equal length, therefore they could be meaned as being equally reliable. The good agreement is due to the very small angular error of only 5 seconds in the Base Traverse.

6.5.5   Correction of the bearings

6.5.5.1   Link traverse

Having computed the *bearings misclosure*, this must be distributed through the bearings of the Gateway Traverse. Normally the bearings error is divided by the number of stations in the traverse, including the terminals (01 and 12), and the result is the angular correction to be applied at each station, with reversed sign, rounded to the nearest second. Application of this to the Gateway Traverse gives:

	Leg	Deduced bearing*	Correction	Corrected bearing	
	01–21	167° 00′ 05″	+19″	167° 00′ 24″	
	21–31	182 13 03	+39	182 13 42	
	31–32	192 59 51	+59	193 00 50	
	32–12	167 14 36	+79	167 15 55	
	12–13	94 01 54	+98	94 03 32	⎫
					⎬ from co-ordinates
or	12–11	185 11 01	+99	185 12 40	⎭

* from Section 6.5.4

Leg 01–21 was corrected by +19″, leg 21–31 by +19+20, leg 31–32 by +19+20+20, and so on, until the correct closing bearings were reached. Note the arbitrary alternation of +19 and +20, to avoid using decimals of a second.

In effect, the application of +19″ on bearing 01–21 changes ALL the bearings by +19, then a further application of +20″ at bearing 21–31 changes that and all subsequent bearings by an additional +20, and so on. The tabular method is the easier to follow in practice.

It should be noted that when a traverse has a mix of long and short legs, it is customary to apply a greater correction at angles observed between two short legs than at angles between two long legs, the former can absorb greater correction than the latter without displacing the traverse.

The methods of correcting bearings or angles before calculating co-ordinates are based on the traditional belief that the angular measurement is more accurate than the linear measurement in a traverse. Where modern high accuracy EDM methods of distance measurement are used, some surveyors consider that, provided the angular misclosure is seen to be acceptable, then the linear measurement is likely to be as accurate as the angles and they do not adjust the angles. Instead, they leave the bearings as observed, use them in the co-ordinate calculations, then adjust the resulting co-ordinates. Whichever methods of adjustment and correction are used, of course, the results should be similar within a given specification.

6.5.5.2 Closed loop traverse

Section 6.5.4.2 illustrated how to work through the bearings of a closed loop traverse where the figure angles had been previously adjusted to the theoretical total of $(2n-4)$ or $(2n+4)$ right angles. It is not essential to adjust the angles first and had this not been done then the method of Section 6.5.5.1 could be used.

TRAVERSE COMPUTATION SHEET (Polar to rectangular)

Traverse .....Gateway.......Calculated by ...K. Drewery.......Date..17 Jun '83.

Total length th.207.601.m..Co-ordinate misclosure .... Accuracy 1/.........

Legs and Stn's	Bearings and Angles	Cn	Corrected Bearings	Distance m	E.	N.	St'n
	POLAR CO-ORDINATES;				RECTANGULAR CO-ORDINATES		
01-02	180 00 00						
+∠01	347 00 05	+			500·000	500·000	01
01-21	167 00 05	19"	167 00 24	39·754	508·938	461·264	
01-17	63 13 08	⎫					
+∠01	103 47 02	⎬ Check only					
01-21	167 00 10	⎭					
21-01	347 00 05						
+∠21	195 12 58	+					
21-31	182 13 03	39"	182 13 42	32·488	507·675	428·800	
31-21	02 13 03						
+∠31	190 46 48	+					
31-32	192 59 51	59"	193 00 50	77·578	490·205	353·215	
32-31	12 59 51						
+∠32	154 14 45	+					
32-12	167 14 36	79"	167 15 55	57·781	502·943	296·855	
12-32	347 14 36						
+∠12	106 47 18	+			502·948	296·843	12
12-13	94 01 54	98"	94 03 32				
12-13 (overleaf)	94 03 32						
Misclosure	−01'38"						
12-32	347 14 36						
+∠12	197 56 25	+					
12-11	185 11 01	99"	185 12 40				
12-11 (overleaf)	185 12 40						
Misclosure	−01'39"						

Figure 6.14

6.5.6   Calculation of rectangular co-ordinates

6.5.6.1   *The traverse computation sheet*

The various calculations involved in a traverse are best done together, on a single computation sheet, even though they have been treated separately up to this point. *Figure 6.14* shows a typical computation sheet, for use with a calculator, with all the polar co-ordinate data for the Gateway Traverse ready for conversion to rectangular co-ordinates. There is no standard form in use, but this version is suited for use with scientific and trigonometric pocket calculators and it does fit on to an A4 sheet, ideal for binding or photocopying.

Note that the computed rectangular co-ordinates have already been entered here, the next section shows how these are calculated.

6.5.6.2   *Co-ordinates and partial co-ordinates*

In *Figure 6.15* the point 0 may be considered to represent a *station*, while the line OP represents an arbitrary *traverse leg* with length equal to unity (i.e. one). The direction OY is north direction and the *bearing* of the line OP is the clockwise angle measured in turning from OY direction to OP.

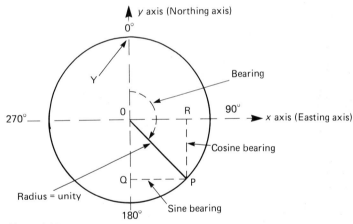

Figure 6.15

If a perpendicular is drawn from P to Q on the *y* or Northing axis, then the distance PQ is equal to the sine of the bearing of OP. Similarly, a perpendicular drawn from P to R on the *x* or Easting axis will give a length PR equal to the cosine of the bearing of OP. In the figure, the sine is positive, since P is to the east of the *y* axis, and the cosine is negative, since P is south of the *x* axis. This is effectively the mathematician's approach turned through 90° and with a mirror image. Surveyors and navigators always consider angles as being measured clockwise from north in this way, and there is no problem with scientific calculators.

If OP represented a real traverse leg, running from station O to station P, with length OP, then the distances PQ and PR are its *partial easting*, or difference easting ($\triangle E$) and *partial northing,* of difference northing ($\triangle N$) respectively, i.e. the partial co-ordinates of line OP. Then

$$\triangle E_{OP} = OP \text{ sin brg OP}$$

and

$$\triangle N_{OP} = OP \text{ cos brg OP}$$

These partial co-ordinates may be calculated in this manner, traditionally with log tables but today using a calculator with sin and cos functions or, preferably, using a calculator with a polar to rectangular key, P→R.

It will be clear that denoting the co-ordinates of point O as E_O, N_O, and of point P as E_P, N_P, then

$$\triangle E_{OP} = E_P - E_O,$$

and

$$\triangle N_{OP} = N_P - N_O,$$

therefore

$$E_P = E_O + OP \text{ sin brg OP},$$

and

$$N_P = N_O + OP \text{ cos brg OP}.$$

In the Gateway Traverse, *Figure 6.12,*

$$E_{21} = E_{01} + (\text{Distance } 01\text{–}21 \times \text{sin brg } 01\text{–}21),$$
$$E_{31} = E_{21} + (\text{Distance } 21\text{–}31 \times \text{sin brg } 21\text{–}31),$$
$$E_{32} = E_{31} + (\text{Distance } 31\text{–}32 \times \text{sin brg } 31\text{–}32), \text{ etc.,}$$

and similarly with the Northing values.

Full advantage of the calculator should be taken by using the store(s) to eliminate the need to write down and re-enter intermediate answers. As an example, the E_{01} and N_{01} values may be keyed into stores 1 and 2 respectively, then the bearing and distance for 01–21 entered and converted to rectangulars using the P→R key. The $\triangle N_{01-21}$ should then be added to store 2 and the $\triangle E_{01-21}$ added to store 1, thus the co-ordinates of station 21 are obtained and these values may be written down on the traverse computation sheet and left in the stores until superseded by the co-ordinates of station 31. The process may be continued in this way until the closing station is reached.

It is important to note that, when using the P→R key, most calculators display Northings before Eastings. The calculated co-ordinates are entered on the computation sheet on the same line as the corrected bearings and distances, as shown in *Figure 6.14,* together with the co-ordinates of the opening and closing stations 01 and 12, on the lines 01 and 12.

The $\triangle E$s and $\triangle N$s used to calculate the co-ordinates may be positive or negative, and most calculators will automatically display and use the correct sign in the calculation of co-ordinates. If this is not so, then it should be remembered that the use of the P→R key or the sine and cosine functions is to determine the $\triangle E$ and $\triangle N$ of a traverse leg. Thus for a traverse leg from station A to station B, it is required to calculate how far station B is east of station A, and how far station B is north of station A. If station B does not lie to the north and east of station B, then one or more of the partial co-ordinates must be negative. The appropriate signs for the quadrants are shown in *Figure 6.16*.

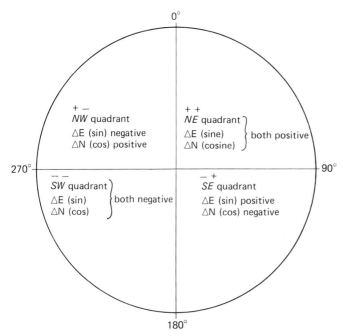

Figure 6.16

6.5.7   Traverse misclosure and accuracy

The final pair of co-ordinates in *Figure 6.14* are the *accepted co-ordinates* of station 12, while the penultimate pair are the co-ordinates of the same station as calculated from the traverse data. In theory, these pairs should be the same and any difference indicates the *co-ordinate misclosure* of the traverse. Subtracting the accepted values from the calculated values gives, in this case, misclosures of −5 mm east and +12 mm north. The *linear misclosure* of the traverse is the actual linear distance between the plan point defined by the accepted co-ordinates and the plan point defined by the calculated co-ordinates. This is equal to the length of the hypotenuse of the right-angled triangle formed by these errors, equal to $(e_E^2 + e_N^2)^{1/2}$ where e_E and e_N are the errors in Eastings and Northings respectively. In this case, 13 mm, as shown in *Figure 6.17*.

Acceptable standards of accuracy are often specified in terms of a stated maximum permissible linear misclosure, such as 50 mm or 200 mm. Another method is to define the accuracy of the traverse as the ratio of the linear misclosure to the total traverse length, in this case 0.013/207.601, which is approximately 1/16000, or 1 in 16000. (Note that in the case of a

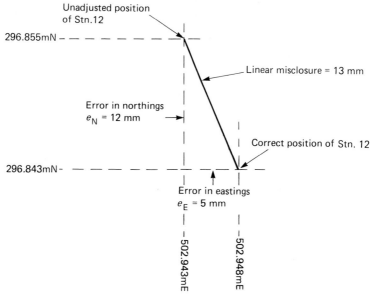

Figure 6.17

closed loop this form is merely an indication of the consistency of the work
and not strictly a measure of absolute accuracy.)

The British Standard referred to previously recommends that traverses
should close to within $\pm 2\sqrt{L}$ mm, where L is the total length of the traverse
in metres. In the Gateway Traverse, this gives 28.8 mm, and the actual
linear misclosure is well within this limit. In specifying permissible linear
misclosures, limits may be variously encountered stated in the forms
K mm, or KL mm, or $K\sqrt{L}$ mm, or $(k + KL)$ mm, where k and K are
suitable factors and L is the traverse length in metres. In the Gateway
Traverse, the technical specification required that the linear misclosure
should not exceed 0.05 m.

6.5.8   Correction to rectangular co-ordinates

If the traverse misclosure exceeds the permissible limit, it may be necessary
to re-measure the traverse and to re-compute. If the accuracy is deemed to
be acceptable, then the co-ordinate misclosures should be distributed
through the traverse, with reversed sign, so that all station co-ordinates are
adjusted and the calculated co-ordinates of the end station match the
accepted co-ordinate values. This process is generally known as 'correction
to co-ords'.

There are a variety of methods of correcting co-ordinates, some perhaps
more mathematically respectable than others, but there is no method
which guarantees that the adjusted co-ordinates are the correct values. A
very widely used rule in practice is *Bowditch's method*, which has no great
theoretical validity but is easy to apply.

In Bowditch's method, each partial co-ordinate for each leg is corrected by an amount which is proportional to the length of the leg. In fact, *the partial easting* for each *leg* is adjusted by

$$\frac{\text{Total misclosure in Eastings}}{\text{Total length of traverse}} \times \text{length of the traverse leg}$$

Similarly, the *partial northing* for each *leg* is adjusted by

$$\frac{\text{Total misclosure in Northings}}{\text{Total length of traverse}} \times \text{length of the traverse leg}$$

Applying this to Gateway Traverse, for each leg:

Correction to $\triangle E = (+0.005/207.601) \times$ length of leg
Correction to $\triangle N = (-0.012/207.601) \times$ length of leg,

then

leg 01–21 (31.754 m) correction $= +0.001$ m E, -0.002 m N
leg 21–31 (32.488 m) correction $= +0.001$ m E, -0.002 m N
leg 31–32 (77.578 m) correction $= +0.002$ m E, -0.005 m N
leg 32–12 (57.781 m) correction $= +0.001$ m E, -0.003 m N

Sum $= +0.005$ m, -0.012 m

Note that in practice, where the error is small as in this example, simple calculation by inspection might well be adequate.

The corrections calculated above are those for the individual partial co-ordinates of the legs, but it is the co-ordinate values which must be adjusted and the corrections must be progressively summed and applied to the co-ordinates. Thus, the corrections to the co-ordinate Eastings are $+0.001$, then $+0.002$ ($+0.001 + 0.001$), then $+0.004$ ($+0.001 + 0.001 + 0.002$), and finally $+0.005$ ($+0.001 + 0.001 + 0.002 + 0.001$).

o rectangular)
i by......Perry.....Date..17.1.13..
isclosure...13 mm....Accuracy 1/16000..

Distance	RECTANGULAR CO-ORDINATES		St'n
	E.	N.	
	500.000	500.000	01
39.754	508.938	461.264	
	+.001	-.002	
	508.939	461.262	21
32.488	507.675	428.800	
	+.002	-.004	
	507.677	428.796	31
77.578	490.205	353.215	
	+.004	-.009	
	490.209	353.206	32
57.781	502.943	296.855	
	+.005	-.012	
	502.948	296.843	12
Misclosure	-.005	+.012	

Figure 6.18a

T (Polar to rectangular)
..Calculated by......Perry.....Date..7 Jun 13
-ordinate misclosure....49 mm....Accuracy 1/4300....

Corrected Bearings.	Distance	RECTANGULAR CO-ORDINATES		St'n
		E.	N.	
		500.000	500.000	01
41.00 05	39.754	508.942	461.265	
		-.009	-.002	
		508.933	461.263	21
52.13 23	32.488	507.685	428.801	
		-.016	-.004	
		507.669	428.797	31
92.59 51	77.578	490.237	353.211	
		-.034	-.009	
		490.203	353.202	32
167.14 36	57.781	502.995	296.856	
		-.047	-.013	
		502.948	296.843	12
Misclosure		+.047	+.013	

Figure 6.18b

Similarly the co-ordinate Northings corrections are −0.002, −0.004, −0.009 and −0.012. The layout of the co-ordinate correction is shown in *Figure 6.18(a)*.

In Section 6.5.5.1 it was suggested that on some occasions it may be considered appropriate NOT to adjust the traverse angles, but rather to use the unadjusted angles in the co-ordinate computation and then apply corrections to the computed co-ordinates only. The effect of this procedure may be observed in *Figure 6.18(b)*, and it will be noted that the maximum difference between the results is 8 mm in the Eastings at station 31, an amount which is negligible in this particular task. Since the traverse runs roughly north to south, it is probable that the misclosure in Eastings is primarily due to angular error.

The calculation of the closed loop traverse co-ordinates, using the same methods, is shown in *Figure 6.19*.

TRAVERSE COMPUTATION SHEET (Polar to rectangular)

Traverse *Necen Hme .+81.*.....Calculated by...*C.Freit.*........Date..*Jly. Yo.*....
Total length.*29.4.:37.m*.Co-ordinate misclosure.....*.73.mm*..Accuracy 1/*4.000*....

Legs and 'Stn's	POLAR CO-ORDINATES				RECTANGULAR CO-ORDINATES		
	Bearings and Angles	Cn	Corrected Bearings	Distance m	E.	N.	St'n
A −D	00° 00						
+∠A	80 27				100.000	100.000	A
A − B	80 27		80°27'	80.30	179.187	113.322	
B − A	260 27				000 +.001	020 −.020	
+∠B	95.51				179.19	113.30	B
B − C	356 18		356' 18	67.85	174.809	181.031	
C − B	176 18				000 +.001	007 −.037	
+∠C	85 54				174.81	180.99	C
C − D	262 12		262°12'	75.51	99.997	170.783	
D − C	82 12				000 +.002	000 −.056	
+∠D	97 48				100.00	170.73	D
D − A	180 00		180' 00'	70.71	99.997	100.073	
A − D	00 00				000 +.003	000 −.073	
(∆₁∩bnt)	00 00				100.00	100.00	A
				Machins	−.003	+.073	

Figure 6.19

6.6 Co-ordinate problems

A wide variety of co-ordinate problems may arise in traversing, and also in other surveying applications, notably in setting out. This section outlines methods of dealing with the common problems.

Section 6.5.6 showed how to calculate the *partial co-ordinates* of a line, given its length and bearing and also how to calculate the *co-ordinates* of the point at one end of a line, given the co-ordinates of the other end point and the line bearing and length. (Equivalent to using the P→R key on a calculator.)

It is often necessary to invert this process and find the *bearing* and *length* of a line from the co-ordinates of its end points. (Equivalent to using the R→P key.) Since for a line AB,

$$\triangle E_{AB} = AB \times \sin \text{brg } AB, \text{ and } \triangle N_{AB} = AB \times \cos \text{brg } AB$$

then

$$\triangle E_{AB}/\triangle N_{AB} = AB \times \sin \text{brg } AB/AB \times \cos \text{brg } AB = \tan \text{brg } AB$$

If the co-ordinates of the end points are given then the partial co-ordinates of the lines may be deduced and the *bearing* is the angle whose tangent equals $\triangle E_{AB}/\triangle N_{AB}$. Thus bearing AB = arc tan $\triangle E_{AB}/\triangle N_{AB}$.

Care must be taken regarding which quadrant is involved, of course, and with the signs. Given the line bearing, then the line length is

$$AB = \triangle E_{AB}/\sin \text{ brg } AB = \triangle N_{AB}/\cos \text{ brg } AB$$

It is often necessary to solve triangles and the useful formulae then are the *sine* and *cosine rules*.

Sine rule

$a/\sin A = b/\sin B = c/\sin C$, or $a = b \sin A/\sin B$, $\sin A = a \sin B/b$.

Cosine rule

$a^2 = b^2 + c^2 - 2 bc \cos A$, or $\cos A = (b^2 + c^2 - a^2)/2 bc$.

6.6.1   Plane rectangular co-ordinate transformation

Where a co-ordinate grid has been superimposed on a site, while a different grid has been used for the original survey, it may be necessary to convert the co-ordinates of points from one grid to the other. In order to carry out such a transformation the co-ordinates of two points must be known, on both grid systems.

Given the two sets of co-ordinates for the common points, then the shift, swing and scale may be calculated. The *shift* is the displacement of the grid origins with respect to one another, in eastings and northings. *Swing* is the angular difference between the two grid norths and *scale* is the ratio between the two lengths of a line common to both grid systems. When these have been calculated then the remaining co-ordinated points on one grid may be transformed to the other grid.

The formulae required are:

$$E''p = e + s(E'p \cos \gamma - N'p \sin \gamma) \ldots$$
$$N''p = n + s(N'p \cos \gamma + E'p \sin \gamma) \ldots$$

where $E'p$ and $N'p$ are the original co-ordinates of a point, $E''p$ and $N''p$ are the transformed co-ordinates of the same point, e and n are the easting and northing shifts respectively, s = scale and γ = swing. (See *Figure 6.20*).

The scale is calculated as $s = d''/d'$, where d' and d'' are the original and transformed distances between two points common to both grid systems. The swing is calculated as

$$\gamma = \beta' - \beta''$$

where β' and β'' are the original and transformed bearings of the line between the same two common points.

The shift constants are calculated from

$$e = E''p - s(E'p \cos \gamma + N'p \sin \gamma) \ldots$$
$$n = N''p - s(N'p \cos \gamma - E'p \sin \gamma) \ldots$$

A more advanced text should be consulted for the derivation of these formulae, which are of particular value in rectangular grid setting out.

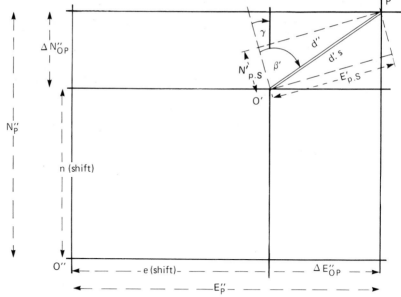

Figure 6.20

6.6.2   Problems in linking to control

6.6.2.1   The 'trilateration' problem

Figure 6.21 illustrates a traverse with stations 1, 2, 3, . . ., to be linked to existing co-ordinated control, it is required to find the opening co-ordinates of station 1 and the opening bearing of the first leg 1–2. Two control points A and B are in the vicinity (the triangle AB1 must be well-conditioned), and the distances 1–A and 1–B and the angle α have been measured, but the theodolite cannot be located over points A or B. Provided 1–A and 1–B are at least as long as the average traverse leg then the triangle AB1 may be solved, the opening co-ordinates of 1 computed and the opening bearing 1–2 deduced.

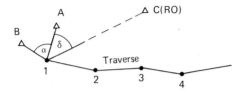

Figure 6.21

If the lines 1–A and 1–B are shorter than the average traverse leg, then an additional angle δ must be observed to a distant RO, a control point C, then the problem may be solved as follows:

(i) Calculate the bearing and length of line A–B from co-ordinates, using R→P.

(ii) Using the cosine rule, calculate the angles A and B in triangle AB1.
(iii) Check that the angles at A, B and 1 in triangle AB1 sum to 180° ±
an acceptable tolerance – do not adjust the angles.
(iv) Apply the calculated angles to the bearing AB to obtain bearings
A1 and B1.
(v) Calculate and check the opening co-ordinates of 1 from A and from
B, by bearing and distance, P→R.
(vi) Calculate the bearing of line 1–C from co-ordinates, using R→P.
(vii) Taking C as the RO, determine the opening bearing of leg 1–2 as
usual.
(viii) Compute the traverse as normal.

6.6.2.2 The 'satellite' problem

Figure 6.22 illustrates a situation with traverse 1, 2, 3, . . . to be linked to a
co-ordinated control point C over which the theodolite cannot be located.
There is a distant control point A which will serve as the RO. The
theodolite is placed at a point S as close as possible to C, forming a
'satellite station' to C. The bearing and length of line C–1 are required in
order to find the co-ordinates of 1 and the bearing of leg 1–2.

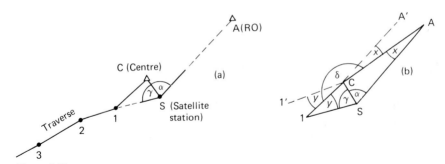

Figure 6.22

To solve this problem, the lines C–1 and C–S are measured, and the
angles α and γ observed, then the angle δ (equal to $\alpha + \gamma + x + y$) may be
deduced. Note that in *Figure 6.22(b)* the pecked lines C–A' and C–1' are
parallel to S–A and S–1 respectively. The procedure then is:

(i) Calculate the bearing and length of the line C–A from co-ordinates,
R→P.
(ii) Using the sine rule in triangles CAS and C1S, determine the angles x
and y respectively.
(iii) Obtain δ in this example $= \alpha + \gamma + x + y$.
(iv) Apply the angle δ to the bearing CA to obtain the bearing C1 and
thus the bearing 1C.
(v) Taking C as the RO calculate the bearing of leg 1–2 as usual.
(vi) Calculate the co-ordinates of 1 from C by bearing and distance,
P→R.
(vii) Compute the traverse as normal.

6.6.2.3   The 'auxiliary station' problem

Figure 6.23 illustrates a similar problem to the last, with a traverse 1, 2, 3, . . . to be linked to control point B which cannot be occupied by the theodolite. An auxiliary station A is located, the distance A–B measured, distant control points C and D will be used as RO1 and RO2, and the angles α, β and γ measured at A. The solution is as follows:

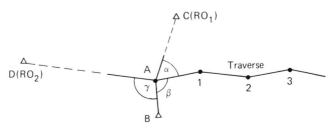

Figure 6.23

(i) Calculate the bearing and length of line B–C from co-ordinates, R→P.
(ii) Using the sine rule in triangle ABC, determine the angle at C and hence deduce the angle at B.
(iii) Apply the angle at B to the bearing BC to obtain the bearing BA.
(iv) Calculate the co-ordinates of A from B by bearing and distance, P→R.
(v) Calculate the bearings of lines A–C and A–D from co-ordinates, R→P.
(vi) Check that the bearing of line C–A from (v) equals the bearing of line C–B from (i) plus the angle at C from (ii).
(vii) Check that the bearing of line A–D from (v) equals the bearing of line A–C from (v) plus angles $\alpha + \beta + \gamma$.
(viii) On the basis of the comparisons made above, decide whether or not to accept the calculated co-ordinates of A. A small change in the length of line A–B, perhaps within its measuring accuracy, would alter the co-ordinates of A and perhaps improve the comparison between the two calculated bearings for the line A–D. Repeat as needed.
(ix) Taking C as RO1 and D as RO2, determine the bearing of line A–1.
(x) Compute the traverse as normal.

6.6.2.4   The 'up-station' problem

Figure 6.24 illustrates survey stations, A, B and C which are to be co-ordinated, such as for the start of a traverse which is to be linked to control. Three distant control points D, U (termed the 'up-station') and E can be observed from B but all are inaccessible to linear measurement. The angles α_1, α_2, α_3, α_4, β and γ have been observed and the distances A–B and B–C measured. The solution is as follows:

(i) Calculate the angles at U in the triangles ABU and BCU.

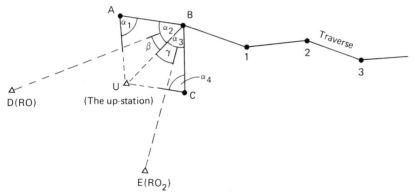

Figure 6.24

(ii) Using the sine rule, calculate the length of the line BU in triangle ABU and in triangle BCU, check that the lengths agree within the accuracy of the linear measurement.

(iii) Calculate the bearings and lengths of the lines U–D and U–E from co-ordinates, R→P.

(iv) Using the sine rule, determine the angles BDU and BEU and hence deduce the angles DUB and EUB.

(v) Apply these angles to the bearings UD and UE and obtain a mean bearing UB.

(vi) Calculate the co-ordinates of B from U by bearing and distance, P→R.

(vii) Compute the traverse as normal.

6.6.2.5   The 'resection' problem

Figure 6.25 illustrates a similar problem to the last – station R (the 'resected point') is to be co-ordinated from angular observations to four existing distant control points A, B, C and D. In *Figure 6.25*, a circle is drawn through the points A, C and R, and this 'danger circle' must not pass through or be close to the point B, otherwise no solution is possible. The

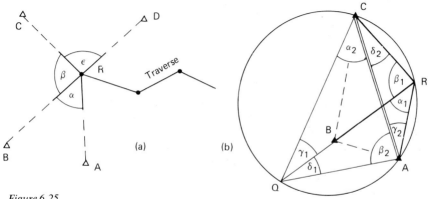

Figure 6.25

line R–B is extended to meet the circumference at Q here, but note that Q might be on the line B–R or on the line B–R extended. The angles α_1 and β_1 are observed and they are equal to the angles α_2 and β_2 respectively. (Angles in the same segment of a circle are equal.)

The 'Collins Point' method of solution is as follows:

(i) Calculate the bearing and length of the line A–C from co-ordinates, R→P.
(ii) Deduce the angle at Q in triangle ACQ ($Q = 180 - \alpha - \beta$).
(iii) Using the sine rule, calculate the lengths of Q–A and Q–C.
(iv) Determine the bearings AQ and CQ.
(v) Calculate the co-ordinates of Q from A by bearing and distance, P→R.
(vi) Calculate the bearing of line Q–B from co-ordinates, R→P.
(vii) Deduce the angles γ_1 and δ_1 from the bearings QA and QB, and QB and QC respectively and hence the angles γ_2 and δ_2.
(viii) Check that $\gamma_1 + \delta_1 =$ angle AQC.
(ix) Using the sine rule in triangle ARC, calculate the lengths of the lines A–R and C–R.
(x) Deduce the bearings AR and CR from the bearing AC calculated in (i) and the angles γ_2 and δ_2.
(xi) Calculate the co-ordinates of R from A by bearing and distance and check from C, P→R.
(xii) Repeat using co-ordinated points ABD, ACD and BCD and obtain a mean value for station R.

6.6.2.6   The 'intersection' problem

Figure 6.26 illustrates this problem. Two co-ordinated points A and B are available, providing a base line A–B whose length and bearing may be determined from co-ordinates in the usual way, R→P. A third point P is to

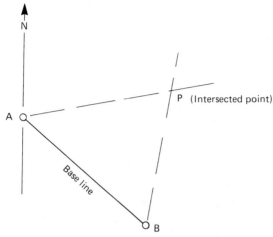

Figure 6.26

be fixed, and this may be done by establishing the rays from A and B through P. The point P may be fixed by either:

(i) observing the angles PAB and PBA, or
(ii) measuring the distances A–P and B–P,

then solving the triangle APB and deducing co-ordinates for P. In either case, a third observation is advisable in order to check the quality of the fix at P. Formula methods are available for this problem but it is probably safer to work from first principles.

6.7   Locating mistakes in traverse calculations

6.7.1   Error in angle

If the bearings have been worked through and the angular misclosure is excessive, the following procedures may help locate an incorrect angle.

(1) Check all arithmetic on the traverse computation sheet.
(2) Check the angles transferred to the computation sheet from the field book/booking sheet.
(3) Check the reductions in the field book/booking sheet, particularly angles close to 90/180°. An obtuse angle may have been entered instead of the reflex, or the FR to the forward station may have been subtracted from the FL to the RO, etc.
(4) If the error is large, say 5° or more, it may be located by either:

(a) roughly plotting the stations on an existing map (if any) then measuring the angles by protractor and comparing with the observed values, or
(b) plot the traverse to some scale, plot the linear misclosure, bisect this line, raise a perpendicular at the bisection. If there is a single gross error the perpendicular will pass through or close to the station in error.

(5) Work through the bearings from each end of the traverse in turn, compute uncorrected co-ordinates for the two traverses. Where the co-ordinates are the same or nearly so may indicate the station in error. If no stations appear to coincide there is more than one angular error.
(6) If no mistake can be found in the calculations, re-observe the station thought to be in error, otherwise re-observe all stations.

6.7.2 Error in co-ordinate calculations

If the angles have been accepted and adjusted, yet the linear misclosure is excessive, the following procedures may help locate the error.

(1) Calculate the bearing and length of the misclosure. Any leg with a similar bearing (or ±180) is suspect and the error in the leg would approximate to the misclosure length. If many legs are all on the same approximate bearing, they may all need to be re-measured.
(2) If the misclosure bearing is not similar to any leg bearing it indicates more than one error or more likely an error in the calculations. Depending upon method of calculation, the error might be:

incorrect data entry to calculator,
sine and cosine interchanged,
+ and − interchanged,
$\triangle E$ and $\triangle N$ interchanged,
incorrect addition or subtraction,
error in reduction of measured lengths,
error in abstraction from field book/booking sheet.

(3) Compare the distances on a rough plot as in (4)(a) above.
(4) Re-measure legs as necessary. Gross errors may only be avoided by the use of suitable working procedures. In the calculations, the best way to detect error is to have independent calculations by two different people, each abstracting the information independently from the field book/booking sheet.

6.8   Traverse plotting

The plotting of a co-ordinated traverse is simple and fast. A grid of mutually perpendicular and parallel lines is drawn on the plan, at intervals of about 200 mm, actual distances depending upon the scale of the plan. The grid should not be drawn by tee-square and set-square, these being highly suspect generally, but with a steel straightedge with scale, beam compasses, a good quality pencil finely sharpened and a suitable piece of drawing material. When lightly drawn-in with fine pencil lines the grid intersections should be marked with blue ink crosses for permanence and co-ordinate values assigned to the lines and marked against each. *Figure 6.27* illustrates the grid construction method.

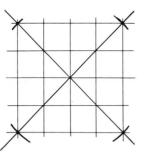

(i)   Draw oblique lines in the approximate position of the diagonals
(ii) From the intersection of the oblique lines describe four arcs of equal radius cutting these lines
(iii) Join the cuts so formed, to form a perfect rectangle
(iv) Check with scale that opposite sides are equal
(v)   Subdivide as necessary and construct squares
(vi) Check at random the lengths of the diagonals of some of the squares and rectangles
(vii) Erase oblique lines

Figure 6.27

The best equipment for grid construction and plotting traverses is the rectangular co-ordinatograph, a precision instrument fitted to a flat rectangular table top. It has a scale along one edge and a second scale at right angles may be moved along the first scale. The second scale carries a microscope or plotting tool such as a pen, pencil or pricker.
    The microscope/plotting tool may be moved over the table surface, the two scales giving easting and northing co-ordinate values for precision

plotting of the traverse stations. Some co-ordinatographs are suitable for attachment to computers and photogrammetric equipment, as mentioned in Chapter 15.

When plotting by hand with a drawn grid, each survey station is plotted as follows:

Each ordinate (Northing) is plotted twice, once on each of the outside ordinates of the relevant grid square containing the point. The straightedge is laid to connect the plotted positions and a visual check made that the straightedge appears parallel to the lateral grid lines. If it appears parallel, a fine pencil line is drawn to connect the plotted positions. The appropriate Easting ordinate is plotted in a similar manner, then the station position marked and circled lightly in pencil, annotating with station name or number. All stations are located on the grid in this way, then the leg lengths scaled off as a check. If all checks, then the station points may be pricked through and any unnecessary construction lines erased.

Detail, if any, is plotted in accordance with Chapters 4, 7 and 14.

6.9   Microcomputer applications

The MICROSURVEY system includes programs for the majority of topics considered in this chapter.

6.9.1   Determination of opening bearing

The Initial Bearing program may be used to determine the opening bearing of a traverse survey, using data for either 1 or 2 Reference Objects. The traverse may be a closed loop (ring), a closed link, or open. The data required include the co-ordinates of the start station and RO(s) and the angle(s) observed between the RO(s) and the forward station. The program computes the forward bearing(s) and if there are two ROs both values are displayed and the user must decide the value to be accepted.

This program provides no data storage, but a hard copy of the data and results may be obtained.

6.9.2   Working through bearings

The Bearings program may be used to determine the adjusted bearings of a link traverse between two points, given the appropriate data in respect of the opening and closing Reference Objects, the first and last stations of the traverse and the included angles measured at each station. The opening and closing bearings are treated in a similar manner to that used in the Initial Bearing program, so that the final decisions as regards these matters and the error to be distributed are made by the user.

The program may not be used with an open traverse, since that has no data on which adjustment may be based. It is not used with a closed loop, since the Traverse program automatically adjusts the bearings of a closed loop.

150   Traverse survey

Figures 6.28 to *6.30* illustrate the printer output from working through the bearings of the Gateway Traverse – in this case the user accepted the bearing determined from opening RO1, and the misclosure as 1′ 38″ and the program adjusted the bearings right through.

All the usual disk file data storage, amendment, printout and display facilities are provided. In addition, however, a file of bearings data may be created which can be read by the Traverse program, thus obviating the need to key angle data for a link traverse into the Traverse program.

```
             Leicester  Polytechnic

             BEARINGS   DATA

Survey bearings data printout from CBM micro program BEARINGS


     Copy 1    Dated 7th August 1984    (Data held on file b-gateway)

     Traverse bearings :  Gateway Traverse

     Client :  R E Paul

     Surveyed by  R Drewery on 17th June 1983


Summary of Bearings and Data Entered


Fixed Co-ordinate List


     Opening points              E            N

     R.O.  1 (Pt 02      )     500.000      125.034
     R.O.  2 (Pt 17      )     533.353      516.834
     Stn.  0 (Pt 01      )     500.000      500.000


     Closing points

     R.O.  1 (Pt 11      )     495.521      215.410
     R.O.  2 (Pt 13      )     619.801      288.551
     Stn.  4 (Pt 12      )     502.948      296.843


Observed Opening Included Angles

     R.O.1 - 0 - 1    347 00 05
     R.O.2 - 0 - 1    103 47 02


Calculated Initial Bearings of Line 0 - 1

     Using R.O.1 -    167 00 05
     Using R.O.2 -    167 00 10


Accepted Initial Bearing of Line 0 - 1

     Bearing  0 - 1   167 00 05


Continued -
```

Figure 6.28

Page 2   Traverse bearings : Gateway Traverse

Observed Traverse Included Angles

```
        0 -  1 -  2      195 12 58
        1 -  2 -  3      190 46 48
        2 -  3 -  4      154 14 45
```

Observed Closing Angles

```
        3 -  4 - RO1     197 56 25
        3 -  4 - RO2     106 47 18
```

Computed Closing Bearings at Stn. 4 from Co-ordinates

```
      To closing R.O.1     -     185 12 40
      To closing R.O.2     -      94 03 32
```

Closing Bearings at Stn. 4 as brought forward

```
      To closing R.O.1     -     185 11 01    Error   -00 01 39
      To closing R.O.2     -      94 01 54    Error   -00 01 38

      Accepted bearing error                          -00 01 38

      Correction to be applied                         00 01 38
```

Continued --

Figure 6.29

Page 3   Traverse bearings : Gateway Traverse

Traverse Bearings Summary

Stn	Name	Line	Bearings List Observed	Adjusted
Start 0	Pt 01	0 - 1	167 00 05	167 00 25
1	Pt 21	1 - 2	182 13 03	182 13 42
2	Pt 31	2 - 3	192 59 51	193 00 50
3	Pt 32	3 - 4	167 14 36	167 15 54
End 4	Pt 12	4 - RO1	185 11 01	185 12 39
4		4 - RO2	94 01 54	94 03 32

Figure 6.30

6.9.3   Reduction of line lengths

As shown in Section 5.4, the Lines program may be used to reduce the lines of a taped traverse and write a file of lines data to be read by the Traverse program.

6.9.4 Traverse computation

The Traverse program can compute the co-ordinates of the stations of a traverse of up to 25 legs, given the appropriate linear and angular measurements and the co-ordinates of the start and end (if any) stations. The traverse may be a closed loop, a closed link, or open, the co-ordinates of closed traverses being automatically adjusted by Bowditch's method.

Figure 6.31 shows the data printout in respect of the Gateway Traverse, *Figures 6.32* and *6.33* show the misclosures/accuracy and the computed co-ordinates respectively. Note that the 'Comparison factor' is that published by Allen, Hollwey and Maynes. The results printout uses the traditional 'Traverse Comp' layout, including partial co-ordinates. The results differ slightly from the hand results, since the bearings are adjusted equally by the program and a different angle error was accepted.

```
              Leicester  Polytechnic

              TRAVERSE  DATA

Traverse survey data printout from CBM micro program TRAVERSE


      Copy 1    Dated 7th August 1984 (Data held on file t-gateway)

      Survey :  Campus survey - Gateway Traverse
                (Link traverse between two points)

      Client :  R E Paul

      Surveyed by  R Drewery on 17th June 1983


Summary of Data as Entered
_____

      Start co-ordinates

      Easting 0       500.000     Northing 0       500.000

      Closing co-ordinates

      Easting 4       502.948     Northing 4       296.843

Station    Name      Adj. Bearing     Line        Length

      0      Pt 01     167 00 25     0 - 1       39.754
      1      Pt 21     182 13 42     1 - 2       32.488
      2      Pt 31     193 00 50     2 - 3       77.578
      3      Pt 32     167 15 54     3 - 4       57.781
      4      Pt 12

Bearings error         -00 01 38     Line Sum    207.601
_____

   ■MICROSURVEY■  Software by Construction Measurement Systems Ltd
      Copyright (C) W.S.Whyte 1981/2/3  Tel: (0537 58) 283
      8000/2.1/8050/28 - for Leicester Polytechnic - 12.1.84
_____
```

Figure 6.31

Leicester Polytechnic

TRAVERSE RESULTS

Traverse computation printout from CBM micro program TRAVERSE

    Copy 1   Dated 7th August 1984 (Data held on file t-gateway)

    Survey :   Campus survey - Gateway Traverse
               (Link traverse between two points)

    Client :   R E Paul

    Surveyed by   R Drewery on 17th June 1983

Misclosures and Accuracy

Co-ordinate Misclosures

    Error in Eastings   = -0.0054

    Error in Northings   = 0.0123

Traverse Accuracy

    Linear misclosure   = 0.0134

    Traverse length   =   207.601

    Proportional error   = 1 in 15448

    Comparison factor Q   = 0.0009

    (Q = lin.misclose/root sum of lines, range precise 0.003/low 0.017)

Angle Misclosure

    Angle misclosure factor X = 49.0000

    (Angle misclosure seconds = X times root of number of lines)

    Angle misclosure (seconds) = 98

    Bearings error   = -00 01 38

Continued -

Figure 6.32

As mentioned, Lines and Bearings program data files may be read directly by the program, thus avoiding the need to enter data twice. The standard disk file data storage, amendment, printout and display facilities are included.

6.9.5   Control link problems

Control link problems (intersection, resection, etc.,) may be handled using the co-ordinate and triangle solution routines in the Trigonometry program.

Page 2:   Survey : Campus survey - Gateway Traverse
                    (Link traverse between two points)

Summary of Final Values

Stn	Line Length	W.C. Bearing o ′ ″	Part. Co-ords. E	N	Co-ordinates E	N
0 (Pt 01)					500.000	500.000
	39.754	167 00 25	8.939	-38.739		
1 (Pt 21)					508.939	461.261
	32.488	182 13 42	-1.262	-32.465		
2 (Pt 31)					507.677	428.796
	77.578	193 00 50	-17.467	-75.590		
3 (Pt 32)					490.209	353.206
	57.781	167 15 54	12.739	-56.363		
4 (Pt 12)					502.948	296.843

MICROSURVEY Software by Construction Measurement Systems Ltd
Copyright (C) W.S.Whyte 1981/2/3   Tel: (0537 58) 283
8000/2.1/8050/28 - for Leicester Polytechnic - 12.1.84

Figure 6.33

Leicester Polytechnic

CO-ORDINATE LIST

Co-ordinate printout from CBM micro program CO-ORDLISTER

Copy 1   Dated 17th June 1983 (Data held on file cl-gateway)

Job name :   Campus survey - Gateway Traverse

Client   :   R E Paul

Surveyor :   R Drewery

Listing of Co-ordinates

Sorted Co-ordinates List

Point	Name	E	N
1	Pt 32	490.209	353.206
2	Pt 01	500.000	500.000
3	Pt 12	502.948	296.843
4	Pt 31	507.677	428.796
5	Pt 21	508.939	461.261

MICROSURVEY Software by Construction Measurement Systems Ltd
Copyright (C) W.S.Whyte 1983   Tel: (0537 58) 283
8000/2.1/8050/28 - for Leicester Polytechnic - 12.1.84

Figure 6.34

6.9.6   Co-ordinate transformation

The Transform program may be used to transform a set of up to 200 point co-ordinates from one plane rectangular co-ordinate system to another, given the co-ordinates of two points in both systems. The basic transformation data are entered at the keyboard, then lists of co-ordinates may be entered by hand or read from data files created by the Traverse, Transform, Co-ordlister or Horizontal Curve programs. The usual disk file data storage, amendment, printout and display facilities are provided.

6.9.7   Co-ordinate list and sort

The Co-ordlister program accepts direct entry of point co-ordinate lists and also reads co-ordinate list files in the same way as Transform. Lists may be concatenated and printed out or displayed sorted or unsorted. (Sorting is on Northings within Eastings.) This facility is particularly useful in bringing together co-ordinate lists produced in, for example, several associated traverses, combining these into a single list for the job. Such combined lists may be edited to add, delete or amend point details, up to a maximum of 400 points.

In *Figure 6.34*, a file of traverse station co-ordinates (tcl-gateway) has been read by Co-ordlister, then sorted to create a new file 'cl-gateway' (Co-ordinate list for Gateway). Several traverses in the actual Campus Survey were concatenated in this way, then common points deleted to provide a final list which was transformed to NG after linking to co-ordinated NG control points.

Chapter 7

Building surveys

7.1 Introduction

Building surveys are usually for the purpose of preparing plans of an existing building which is to be altered – record plans are seldom available and when they do exist they are generally inaccurate. It may also be required to prepare a plan of a very old building for conservation records. A survey should provide all information necessary to prepare not merely a ground plan but also a plan of each floor in the building, together with measured *elevations* (external views) and *sections* ('cut-away' views) of the buildings. While in land survey the surveyor may have to describe boundary features, vegetation and street furniture (post boxes, etc.), in the case of buildings he may have to describe the constructional details and the building services.

The essential difference between these surveys and the chain surveys described in Chapter 4, lies in the plotting scales used. Chain survey scales are typically 1:5000 to 1:500 and building survey scales 1:100 or 1:50. The larger scales are needed in order to communicate the information adequately to the client. Since this may require measurements accurate to the nearest 10 mm, and occasionally 5 mm, the chain is not suitable and offsets may become unacceptable unless over very short distances of a metre or less. It is not difficult to keep lines straight generally, since the lengths of the lines to be measured are often under one tape length. However, the network of lines required for the framework and the techniques used are very similar to chain survey, except in the booking. A small task could be sketched in detail on one sheet of paper, then all the measurements taken and noted on that same sheet. It is not always essential to letter each station, since they are all on the same sheet and confusion is unlikely to arise. A larger survey might require two or more sheets of sketches and in this case it may be helpful to identify points by letters – it is also important to ensure that the several sketches are adequately connected by common points and lines of detail and do not become a series of unconnected parts.

The ground plan of the building is often required in its relationship to the site as a whole; this means that a *site plan* or *plot survey* will be necessary.

This may be at a different scale from the building drawings. If the site plan is to be at a scale of 1:500 then chain survey techniques may be used. At smaller scales a tracing from the National or Municipal large-scale maps of the site may be acceptable, but this may involve copyright approval and the payment of fees to the mapping agency.

At larger scales, chain survey methods are unacceptable for plot surveys, due to the increased time and effort needed to achieve the desired accuracy. Alternative techniques include intersection or bearing and distance, as referred to in Chapters 1 and 14. Conversion of the field observations to rectangular co-ordinates may simplify the plotting. Spot levels or contours may be required over a site, but these are considered in Chapter 8.

On occasion, surveyors may be requested to measure the movement in buildings and other structures. Techniques for this are described at the end of this chapter.

7.2   The plot survey

7.2.1   Typical equipment

The typical equipment is as for chain survey, but a steel tape is to be preferred to a chain. Additional equipment includes

2 m or 3 m retractable steel tape rule,
2 m folding rod,
builder's line and plumb-bob,
manhole key,

and possibly

hand lamp or torch,
theodolite,
electromagnetic distance measuring equipment.

7.2.2   Field procedure

7.2.2.1   Preliminaries

Check the exact purpose of the survey, e.g. simple site plan, site and building plan with drainage layout, or a full building survey, and the scale expected. The survey provides information for the plan, but the client may require details as to the services available, future development in the area, planning restrictions, building lines, etc. It may be necessary to contact the local authority for the area, gas board, electricity board, telecommunication services and other interested parties.

Arrange access to the site and buildings and all permissions.

Check equipment, standardize tapes as needed.

Obtain a sketch map of the site, or an outdated plan, or even a plan at a smaller scale, to provide a rough guide in reconnaissance and planning line layout.

7.2.2.2   Site reconnaissance

A reconnaissance must be carried out, as described in Chapter 4, to fix the positions of lines. Typically the lines should break the area in triangles, each triangle having a check and the triangles tied to one long base line through the site if at all possible. The base line is of particular importance on long, narrow sites and it may be advisable to consider the use of a theodolite to establish the line, especially if the ground is covered with vegetation. Offsets will generally be few since it is often possible to run lines along straight boundary features. On occasion lines may have to be run through the building(s) and in such cases it is helpful if the lines are either parallel or perpendicular to the building faces or walls.

Where the plotting scale is to be 1:200 or greater, or if the site is long and narrow, it may be more effective to use bearing and distance techniques as described in Chapters 12 to 14. Where the site is small and no assistant is available, intersection techniques may be suitable.

7.2.2.3   Linear measurement (and possibly angular measurement)

The methods to be used for direct linear measurement and angular measurement have been covered in the preceding chapters. For *electromagnetic distance measurement,* see Chapter 14.

7.2.2.4   Booking the measurements

When the site is small and there is only a small number of offsets, the booking may be carried out on a single sheet of paper, with single (skeleton) lines representing the chain lines. If the site is large, perhaps with many offsets, it may be better to use the conventional chain survey booking methods. The general guidelines of Chapter 4 are relevant, whichever style is used, particularly as regards clarity and accuracy.

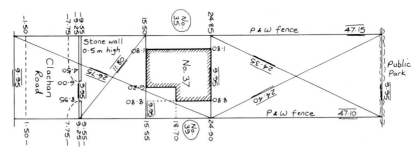

Figure 7.1

Figure 7.1 illustrates a 'single page' survey booking. Note that the techniques of chainage, offsets and straights are similar, including the use of running measurements and overall line lengths to maintain accuracy and avoid ambiguity. Site plans may also include the outlines of neighbouring buildings, footways, roads, sewers, drains, lamp-posts, telegraph poles, etc. The booking of bearing and distance measurements are described in Chapter 14.

7.3   The building survey

7.3.1   Equipment

The equipment is generally as identified in Section 7.2.1, but in addition the following may be needed

sectional ladder,
binoculars,
surveyor's level and levelling staff.

7.3.2   Field procedure

7.3.2.1   Preliminaries

Check exactly what information is to be provided by the survey and obtain any existing plans, sections, elevations. Obtain keys for all unoccupied buildings and any necessary entry permissions. Check that all equipment needed is available and in good order.

7.3.2.2   Reconnaissance

A thorough 'recce' of the building must be carried out in order to assess the size and nature of the task and decide how best to tackle it. A building is usually a series of blocks, of different shapes and sizes at different levels and the inexperienced surveyor must avoid the temptation to measure them separately and then fit them together like a '3-D' jigsaw puzzle. The survey principles of Section 1.7 must be remembered, particularly 'working from the whole to the part' and the need for the independent check. However, it is always difficult to avoid the fragmentation of the measurements and when this occurs reference to the control framework must be made.

It is often difficult to build-up satisfactory triangles, but every possible diagonal measurement should be taken within rooms and every wall must be measured and, taken together, these can form the required triangular frameworks. Buildings or rooms must never be assumed to be rectangular and the fact that two diagonals of a notionally rectangular room are equal does not mean that the room is rectangular. Again, walls may vary in thickness throughout their length, particularly at intersections with cross walls and partitions and often such changes are detected only by very careful measurement and checking. When dealing with large, complex and irregular shaped buildings it may be advisable to provide a theodolite control traverse.

7.3.2.3   Sketching

Since the building shape is generally small and approximately rectangular on plan, the lines to be measured are usually run along the features to be surveyed, the sketches being drawn first, the bookings (measurement notes) being added later. (This is the opposite of the methods used in chain survey.) These freehand sketches should be as large as possible, preferably on A4 plain or 5 mm square ruled paper and roughly true to shape, but no attempt should be made to sketch to scale – the most important consideration is that all details are clear and the required measurements can be shown without crowding or ambiguity.

Sketches are required as follows

(i) Plans of each floor of the building, including any basement. A plan within the roof space and/or a plan of the roof may also be needed.

Floor plans are drawn in the same way as external sites, that is to say looking down on the floor being drawn, but as if the building had been sliced through horizontally at about 1 m above the level of the floor concerned. It may be possible to trace the ground floor plan sketch as a basis of the detailed sketches for other floors.

Details to be shown include wall openings (doors, windows, hatches, etc.), changes in wall thickness, changes in floor levels (steps, stairs, ramps, etc.) and similar construction detail, with information as to materials, span directions, etc. Services installations and fixtures and fittings (gas, heating, water, electricity, ventilation, power, lighting, communication, cooking, sanitation, washing, etc.) and drainage layouts are all required.

Roof plans include gutters, direction of falls, covering materials, parapets, skylights, stacks, rainwater heads, etc.

(ii) Elevations of the exterior of the building. Elevation sketches need not be completed in their entirety where detail is repetitive. Detail to be shown is as for plans, but also includes sills and lintels, external piping and any detail which is visible on elevation but cannot be shown on plans.

(iii) Sections drawn to show what would be visible if the building was cut through vertically. A section need not be in one straight line across the building in plan (or vertically), it may be stepped if it is more appropriate for the job. In some such cases the section actually becomes a combined section and elevation in effect.

Details shown on sections will be similar to those shown on plans and elevations. The section lines should be drawn on the plan sketches.

As in chain survey, the representation of detail by point, line or area symbols and by annotation by abbreviations on the sketches should be similar to that required on the plotted plans, elevations and sections. Some common abbreviations are listed here. Where relevant, the published recommendations of professional institutions and other bodies, such as the British Standards Institution, should be followed.

Common abbreviations are

agg	aggregate	m	metre
asp	asphalt	MH	manhole
bit	bitumen	mm	millimetre
BM	benchmark	MS	mild steel
bwk	brickwork	PI	petrol interceptor
CI	cast iron	RC	reinforced concrete
conc	concrete	RE	rodding eye
FFL	finished floor level	RWP	rainwater pipe
G	gulley	T&G	tongued and grooved
GL	ground level	TBM	temporary bench mark
GT	grease trap	VP	vent pipe
LP	lamp-post	WI	wrought iron

Figures 7.2 to *7.4* show examples of the representation of detail. Where detail is too small to show up clearly on the main sketch, larger detail sketches should be made on a separate sheet of paper, suitably referenced so that they may be referred to later.

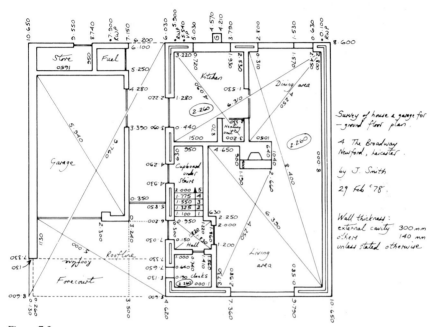

Figure 7.2

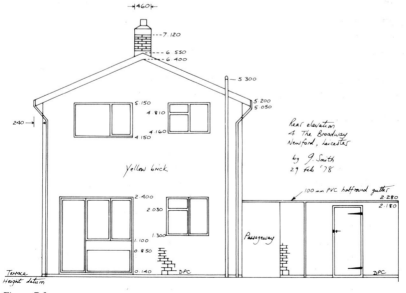

Figure 7.3

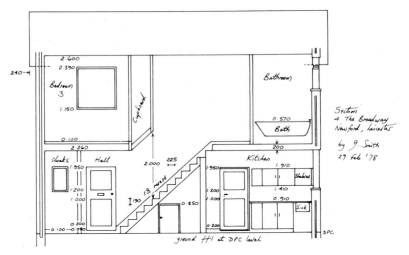

Figure 7.4

7.3.2.4 System of measurement

External linear measurement may be carried out at ground level as described for chaining in Chapter 4, but due to the need to read the tape with a greater accuracy than is demanded in chaining, taping is often carried out with the tape held at chest height and making use of the hook attached to the zero end of the tape.

It is conventional to take all measurements from left to right, i.e. anti-clockwise around the outside of the building and clockwise around the interior of rooms, since in this way the figures on the tape will be the right way up and errors in reading will be minimized. In addition, when plotting later from the bookings this method reduces possible confusion as to which reading is which.

For ease and accuracy in plotting, all measurements along one straight line should be booked as one set of running measurements, provided that the accuracy of measurement can be maintained.

When measuring door openings in internal walls, the measurement along the wall is made to the edge of the actual door, all architraves, facings and so on being ignored. Wall thickness must be measured at every opening in internal or external walls – in this way an unnoticed change of thickness may be detected. Wall thickness should be measured as the actual total thickness of materials – rendering, brick, plaster and all. The actual thickness should be noted, and standard brick sizes should not be assumed.

External vertical distances must be related to a selected suitable datum surface on the building, such as the line of a visible damp-proof course, a plinth or string course, etc. If there is any doubt as to the verticality of a building face it should be checked as in Section 7.5.1. Where it is considered that the selected datum surface is not horizontal, this may be checked using the surveyor's level and staff. When the level is set up, the same staff reading should be obtained from all points on the datum surface if it is level (see Chapter 8). Wherever possible, internal vertical measurements should be linked to the external heights at wall openings.

Inaccessible distances, horizontal or vertical, can often be obtained by counting the number of brick lengths or courses involved then directly measuring a similar length of accessible brickwork. It must be emphasized that this practice should only be used where it is not possible to gain access to the feature to be measured.

The 2 m rod is used to measure heights, but the tape may be dropped for very long or overall heights.

7.3.2.5   *Booking the measurements*

Considerable licence is allowed in the recording of measurements in this type of work, since it is usually plotted by the surveyor himself, but it should be remembered that on occasion others may have to plot the work, hence clarity is essential and the generally accepted rules should be followed.

The methods commonly adopted are as follows

(i) Single or skeleton line booking is used, the lines themselves often not being shown except for tie and check lines, since the lines are represented by the face of the detail, typically wall faces.

(ii) Letters and numbers are not used to identify the terminal points of the lines which are being measured.

(iii) The zero point of a line is entered as $\oplus$ or $\ominus$ if there is any likelihood of confusion as to the position of the measurement zero.

(iv) Running measurements are used for the 'chainage' measurements, and they are written in the direction of measurement.

(v) The length of the line is normally written as for chainage figures, it is only on the tie and check lines (i.e. the diagonals) that the line length is written with the base of the figures on the measured line.

(vi) Offsets, running offsets and plus measurements may be written as in conventional chain survey booking and may be entered either right or left of the line as space permits, provided there is no ambiguity. Occasionally running offsets are also written as short 'chain' lines, but not tied out. To help minimize ambiguity, chainage figures may be written as metres and decimals of a metre, while offsets and plus measurements can be written simply as the number of millimetres, i.e. no decimal point shown.

(vii) Floor-to-ceiling heights should be written in the centre of the floor plan of the area concerned, the figures being encircled to indicate that they are not horizontal dimensions.

Figures 7.2, 7.3 and *7.4* show example bookings for a floor plan, an elevation and a section, using these methods.

7.4   Office procedure – plotting

Plotting consists of several separate plots, including site plan, floor plans, elevations and sections and, depending upon the finished size of these, more than one sheet of drawing material may be needed. The site plan will often be at 1:500, while the remainder will be at 1:100 or 1:50 and occasionally large details at 1:10 or 1:5 will be needed.

7.4.1   Draughting equipment and materials

This is generally as chain survey, with the addition of

drawing board plus tee-square,
set squares, 45° and 60°,
(adjustable set square is useful in addition, but is no substitute for the two simple squares),
protractor,
150 mm compasses/dividers,
springbow compasses/dividers,
steel pricker.

For the alteration or conversion of an existing building, a heavy-grade tracing paper may be suitable. Some individuals, however, prefer to carry out all original plotting on cartridge paper. If a strong, durable and dimensionally stable material is necessary, a modern polyester-based translucent film may be used.

A measured survey of a small building and its site should generally plot on a single sheet of one of the recommended A-size drawing sheets, possibly A2 or A1, without being cramped.

7.4.2   Layout of the survey on the drawing material

The ground floor plan is generally placed at the bottom left-hand corner of the sheet, with the front of the building towards the bottom of the sheet. The first floor plan is then placed alongside the ground floor plan with the same aspect or orientation and the other floor plans similarly.

The front elevation of the building should be drawn immediately above the ground floor plan, again the other elevations are then drawn across the sheet in a row and level with the front elevation. Finally, the sections and site plan should be placed wherever they will conveniently fit into the general arrangement, although the sections should, for preference, be placed alongside the elevations.

Wherever possible, it is important to arrange that common lines on plans, sections and elevations should lie on extensions of their respective lines on the drawing, see *Figure 7.5*. This presentation conforms closely to the 'first angle (or European)' projection, and it is the projection recommended in BS 1192:1969, *Recommendations for Building Drawing Practice*.

Variations are inevitable, thus, for example, a small, single-storey extension at the rear of a building might not require a front elevation to be shown and in each case the surveyor must decide the layout applicable in the circumstances. In all cases the site plan should be drawn with north towards the top of the sheet. BS 1192 also contains recommendations regarding marginal and border information.

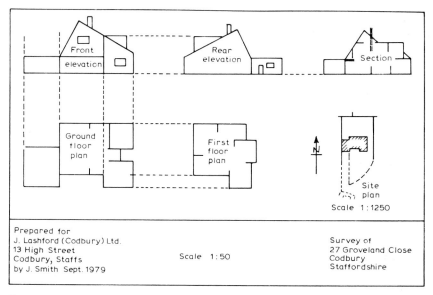

Figure 7.5

The final consideration regarding the layout is that the whole presentation should be well-balanced and pleasing to the eye, as with all drawings.

7.4.3   Plotting the plans

The site plan is plotted like a chain survey, as covered in Chapter 4. (If intersection or bearing and distance techniques are also used, refer to Chapter 14.) The building itself is plotted in a similar manner, though the process is rather more complex. The procedure is as follows.

(1) Draw the longest external wall to scale, in the preferred location on the drawing material.
(2) Using the measured wall thickness, plot the alignment (but not the detail) of the interior face of the same wall. Remember the possibility of different thicknesses along the length of the wall, and make allowance as needed.
(3) On the external wall face, plot the position of all wall openings and raise right angles from these to cut the alignment of the inner face of the wall. Note that there may be rebated or splayed jambs and the construction on either side of an opening may differ; due allowance must be made for such details.
(4) Using the internal wall face measurements, scale off and mark the position of all walls joining into the external wall.
(5) Using the conventional chain survey technique of plotting triangles by swinging arcs, plot the lines representing the faces of all the internal walls of the rooms adjoining the external wall already plotted, using wall lengths and room diagonals to build up the triangles.

(6) Repeat the process until all the ground floor walls have been plotted, applying all possible checks (matching internal and external measurements at openings, checking overall external wall lengths, etc.) since errors can build up rapidly when a building is plotted in this manner.

(7) When the ground floor plan framework is complete and considered to be correct, plot all the detail involved.

(8) Plot the first floor plan by using the ground floor external dimensions (adjusted where necessary) and projecting across from the ground floor plan by the use of T-square and set square. Alternatively, trace the outline of the ground floor plan carefully, then locate the tracing paper where the first floor plan is to be plotted and 'prick' through the corners of the building on to the drawing material. The pricker marks may be joined in pencil to form the outline of the floor plan.

(9) Plot the interior of the first floor plan, in a manner similar to that used for the ground floor plan.

(10) Plot the remaining floor plans in the same way, adapting as necessary if a roof plan is to be drawn.

7.4.4   Plotting the elevations

The procedure to plot the front elevation is as follows.

(1) Project the lines of the external walls and openings of the front of the building upwards from the ground floor plan, as in *Figure 7.5*.

(2) Select an appropriate location and draw a line to represent the chosen datum line for heights, cutting across the lines projected up from the ground floor plan.

(3) Plot all the measured heights on the elevation, above or below the datum line as required, then complete the elevation drawing to show all the field sketch detail.

(4) Check the elevation, correct as necessary.

The remaining elevations should be plotted by a similar combination of projecting, tracing and scaling, as practicable.

7.4.5   Plotting the sections

For preference, the sections should be plotted alongside the elevations, allowing heights to be projected from elevations to sections and reducing the amount of scaling required. The sections are basically plotted in a similar manner to the elevations.

7.4.6   Completing the drawings

The section on completion of chain survey plots, in Chapter 4, is equally relevant to building surveys. Significant differences, however, include a probably greater use of graphical symbols, some or which are illustrated in *Figure 7.6*. There will also be a greater use of descriptive names, abbreviations, annotations and numbers, and it may be necessary to show certain horizontal dimensions and heights.

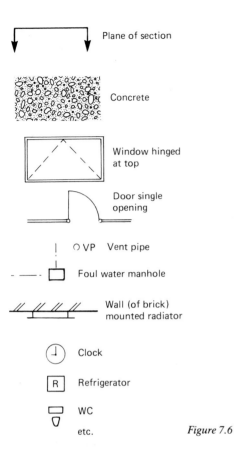

Plane of section

Concrete

Window hinged at top

Door single opening

○ VP   Vent pipe

Foul water manhole

Wall (of brick) mounted radiator

Clock

R   Refrigerator

WC

etc.   *Figure 7.6*

Examples

wall thicknesses,
timber sizes,
materials,
unusual structural details,
floor levels and changes of level,
ceiling heights,
direction of stair rise, number of risers, going,
window and door types,
services information – rising main, meters, etc.,
principal dimensions,
North Point or Points,
scale(s) of the drawing as representative fraction,
title and address of property,
information as to surveyor, when surveyed, etc.

With regard to stairs, each tread in a flight should be numbered in succession, 1, 2, 3, . . ., commencing from the lowest tread. The up direction of the stairs should be shown by an arrow and the word 'up'.

The North Point should show the approximate direction of north, in order to indicate the aspect of the building. It may be placed on the site plan, or alongside the ground floor plan.

The principal dimensions, one length and one breadth, should be shown in each room of each floor plan. A simple arrowed line with a figured dimension is sufficient. The typical floor-to-ceiling height may be shown on one of the sections by a dimension line.

7.5 Building movement

Buildings and other manmade structures are subject to continual movement and when such movement becomes excessive and causes distortion of the fabric (e.g. leaning, bulging and cracking walls), a surveyor may be called in to monitor and record the movement. Movement in structures is recorded as angular or linear changes of position of two or more reference points on the structure with respect to one another and to external points or an external baseline. Measurements for this purpose are sometimes termed 'deformation measurements' and the movements determined may be mathematically converted to measurements on a three-dimensional x, y and z co-ordinate system. The following sections describe possible survey techniques.

7.5.1 Leaning walls and buildings

7.5.1.1 Methods of measuring the amount a wall is 'out of plumb', or its departure from the vertical

(i) By plumb-bob and string with a scale rule. A plumb-bob may be suspended from a projecting wall face, then, when the bob has stabilized, the horizontal distance from the string to the wall face may be measured. This is most suited to short vertical distances and is unreliable in windy conditions. On very short lengths it may be possible to use a bricklayer's spirit level, but this is difficult.

(ii) By optical plumbing techniques. This involves marking a point at high level on the wall, then marking another point on the ground, vertically below the first point. Again, the horizontal distance from the ground point to the face of the wall may be measured. The ground point is located by the vertical alignment control techniques described in Chapter 17.

(iii) By triangulation from a measured base. This entails marking survey points on the structure and establishing a survey base line near the building in such a position that the base line and the pairs of directions from its ends to each survey point form well-conditioned triangles in plan. The base length must be carefully measured, the difference in height of its two end-points determined by levelling (Chapter 8) and the base length reduced to the horizontal. If arbitrary co-ordinates are allotted to one end of the base line, and an arbitrary bearing to its direction, then three-dimensional co-ordinates may be calculated for both ends of the base line.

If, for each survey mark on the building, the horizontal angles to the mark from each end of the base line are measured, then the triangles so formed may be solved and the plan co-ordinates of the marks calculated. Measurement of the vertical angles to each mark from the base line ends, together with the calculated horizontal distances from base line ends to the marks, will allow calculation of the heights of the marks.

The horizontal and vertical angles must be measured by theodolite and angles correct to one second of arc may be required. With a one-second optical-reading theodolite this will entail at least eight measures of each horizontal angle, that is face left and face right on four zeros. In 1983 Wild Heerbrugg of Switzerland introduced the Wild T2000, a one-second electronic theodolite for which they claim that the standard deviation of the mean of a face left and a face right pointing is 0.5" of arc, and it is recommended for deformation measurements.

(iv) By photogrammetric techniques. Photogrammetry is a technique for determining measurements from photographs, described briefly in Chapter 15. The equipment is expensive, but in some very old buildings or large and complex structures, it is the only practicable method.

7.5.1.2   Methods of monitoring movement (if any) in leaning walls and buildings

In monitoring movements in structures, it is necessary to take measurements of the types described immediately above, and to repeat these at intervals of time and compare the successive sets of results. This will usually entail setting up semi-permanent instrument stations well clear of the structure in such positions that they will not, themselves, be affected by movement of the structure.

(i) By the methods in sections 7.5.1.1.

(ii) By setting a theodolite on a fixed line of sight and observing the movement of scales fixed to the building. If horizontal and vertical graduated scales are fixed to the building and the theodolite can be directed in exactly the same line on each visit, variation in the scale readings against the theodolite cross-hairs will demonstrate movement of the structure.

(iii) By electromagnetic distance measurements. If EDM reflectors are placed on survey marks on the structure and a base line established, as in (iii) above, then highly accurate measurements of the distances from the base line terminals to the survey marks may be made and the three-dimensional co-ordinates computed by trilateration methods. The Mekometer by Kern of Switzerland, currently widely considered to be the most accurate EDM instrument, is used in this way for the measurement of dam deformations. Dam deformation measurements can also, of course, be made by the triangulation methods described above.

7.5.2   Bulging walls

Most of the methods of Section 7.5.1 are appropriate, but for short distances (e.g. measurements between wall faces) catenary taping is a possibility. If ordinary catenary taping is not of sufficient accuracy, specialized equipment is available such as the Kern 'distometer ISETH', suitable for lengths from 1 to 50 m. This instrument displays the correct tension and the changes in length on dial gauges. Catenary taping may be useful inside buildings, trenches, tunnels and vaults.

7.5.3   Cracks in walls

To check on the movement at cracks, reference 'tell tale' marks may be placed on either side of the crack and changes in their relative positions over a period of time can indicate linear or linear and angular movement. The marks may be centre punched rivets or screws, with distances between them measured by a rule or vernier calipers. (At least three marks are needed to check on angular movements.)

The 'Avongard Tell-tale', a proprietary device, consists of two overlapping rectangular plastic plates, with suitable scale markings, when fixed over the crack, each plate moves independently of the other.

7.5.4   Settlement and subsidence

Small vertical movements may be monitored by some of the methods detailed above. Large vertical movements, as in mining subsidence, may be measured by levelling as described in Chapter 8, from an area known to be stable, at suitable intervals of time. Large movements such as the movement of cliff faces may be monitored using a network of traverse stations at accessible points at the top, on the face and at the bottom of the cliff. Such traverses are best measured using EDM equipment, the measurements and computing of the three-dimensional co-ordinates of the stations being repeated at intervals.

Chapter 8

Levelling

8.1 Definitions

Levelling is the name given to the method of determining heights by the use of the instrument known as the *surveyor's level*. A *level line* is one which is of constant or uniform height relative to mean sea level and is therefore a curved line concentric with the mean surface of the earth. More formally, a *level line* is a line which lies on one level surface and is normal (at right angles) to the direction of gravity at all points in its length.

A *horizontal line* through a point is tangential to the level line passing through the same point and is normal to the direction of gravity at that point. The difference between a horizontal and a level line passing through the same point must be appreciated. The greater the distance from the common point, the greater the discrepancy. In ordinary levelling, with sights less than 100 m or so, the difference is negligible for practical purposes and may be ignored, see *Figure 8.1*.

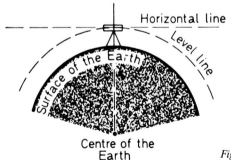

Figure 8.1

A *datum surface* or *datum line* is a level surface or line from which heights are measured, or to which heights may be referred.

A *height* is the vertical distance of a point above or below a datum surface.

A *reduced level* (RL) is the calculated height of a point above or below a datum, as deduced from the surveyor's field observations.

Mean sea level (MSL) is the mean level of the sea as determined at some selected place from observations over a period of time, used as a datum surface for levelling work. The concept is used by many national mapping organizations. *Ordnance Datum* (OD) is the current datum for heights used by the O.S. of Great Britain. It is based on the mean level of the sea at Newlyn in Cornwall, calculated from hourly tide gauge readings recorded between 1915 and 1921.

AOD, seen occasionally on a plan or map after a height value, means *height above Ordnance Datum*.

Bench marks (BMs) are fixed points whose heights relative to a datum surface have been determined using a surveyor's level.

An *Ordnance bench mark* (OBM) is a bench mark established by the O.S., the height of the bench mark relative to Ordnance datum being known accurately. The O.S. have established OBMs all over the mainland and inshore islands of Great Britain, and levelling operations may be referred to these known points.

A *temporary bench mark* (TBM) is a bench mark set up by a surveyor for his own use on a particular job. The TBM height may be established from an OBM, then levels on site may be referred back to the TBM without checking back to the OBM every time. TBMs should be stable, semi-permanent marks, such as concreted pegs or features on a permanent building.

8.2   Ordnance bench marks

The O.S. having established bench marks all over the country, any levelling operations may be referred to an OBM of known height above mean sea level. The O.S. levelling consisted first of lines of primary geodetic levelling, then secondary levelling between these, and finally 'fill-in' by tertiary levelling. The work was originally based on a mean sea level determined at Liverpool in 1844, but this was considered to be unreliable and was superseded by the Newlyn datum for the second and third sets of geodetic levelling. Slight differences appear in the heights of some OBMs shown on O.S. map sheets, depending upon the date of the particular sheet. Post-1956 levels are based on the third geodetic levelling.

8.2.1   Density of OBMs

OBMs are provided to meet normal user requirements and the density of provision varies from under 300 m apart in city areas to over 1000 m apart in rural areas.

8.2.2   Types of OBM

Six different types of BM have been set up by the O.S., similar marks being found in other parts of the world.

Cut bench marks, the commonest form, consist of a horizontal line incised in a vertical surface such as a brick or stone wall. The traditional government 'broad arrow' is cut below the centre of the horizontal line, point upwards, see *Figure 8.2(a)*.

Fundamental bench marks (FBMs) are marks placed on solid rock, at points roughly 40 km apart, and they provide control for the whole of the O.S. levelling network. The mark consists of three reference points, two of which are in a buried chamber not accessible to the public. The third mark is a brass or gunmetal bolt set on top of a low granite or concrete pillar, available for public use.

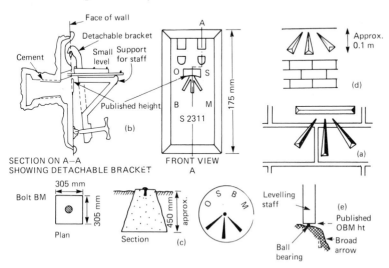

Figure 8.2

Bronze flush bracket OBMs, of lower order than FBMs, are levelling control points set into the side of large public buildings (churches, etc.) and O.S. triangulation pillars. The published height is to the small horizontal platform at the point of the broad arrow marked on the face of the bench mark (*Figure 8.2(b)*). A special staff support (*hanging bracket*) should be used with this type of bench mark. *Brass* or *gunmetal bolts* are used as an alternative to the flush bracket OBMs, where the structure provides no suitable site for the flush bracket. These are 50 or 60 mm diameter mushroom-headed bolts set in solid rock or concrete (*Figure 8.2(c)*).

Brass rivets are an occasional alternative to the standard cut OBM, used where the bench mark must be located on a horizontal surface. If possible, a broad arrow is cut alongside (*Figure 8.2(d)*).

Pivot bench marks are used on horizontal surfaces such as soft sandstone, where the insertion of a rivet would break away the stone. Such an OBM actually consists of a small hole or depression cut to take a pivot, a steel ball bearing of ⅝ in (approximately 16 mm) diameter. In use, the pivot is placed in the depression and the staff held on top of the pivot. The published height refers to the top of the ball bearing (*Figure 8.2(e)*).

8.2.3   Sources of OBM information

O.S. 1:1250 and 1:2500 scale maps and many large-scale plans produced by private survey companies and municipal authorities, indicate bench marks by means of a broad arrow, the head indicating the plan position. The height is usually given and, occasionally, the type of mark. O.S. maps do not always contain the latest available levelling information. In addition, maps may not identify the date of the levelling, the type of mark, or its height above ground level. The O.S. have therefore compiled *OBM lists* which may be purchased direct from the O.S. by any surveyor who needs complete and up-to-date levelling information.

A *bench mark list* is published for each square kilometre, covering the same area which is, or would be, represented on a 1:2500 O.S. map. Both the map and the OBM list are identified by the same national grid reference number. For each bench mark, the list gives the following information

a brief description of the mark, with its location,
the full 10 m national grid reference,
the height of the mark, in metres and feet, with respect to Ordnance datum,
the vertical distance of the mark above ground level, in metres and feet,
the date of the levelling.

When using O.S. maps, a check should be made that the heights shown are on the Newlyn datum. If old maps are used, they may be based on the Liverpool datum and the difference between levels on the two datums is not constant throughout the country.

8.3   Types of levelling

There are two basic types of levelling, *precise* (or *geodetic*) and *ordinary* (or *simple*) *levelling*. The latter is often simply termed *levelling*.

8.3.1   Precise or geodetic levelling

This is the highest order of levelling work, with readings generally observed and recorded to decimals of a millimetre. This form is used for the basic levelling framework of a country, such as the establishment of fundamental bench marks. It is outside the scope of this book, but some information has been provided since it may be necessary to use the techniques in work such as, for example, checking on the movement of structures, certain engineering works and irrigation schemes over very flat areas.

8.3.2   Ordinary or simple levelling

This term covers all levelling work which is not regarded as being precise levelling, with readings taken, at best, to 1 mm. Ordinary levelling may be categorized by its purpose or use, e.g. section levelling, area levelling, construction levelling, etc.

8.4   The principle of levelling

In all forms of levelling, the typical problem is that the height of one point above datum is known and it is required to find the reduced levels of other points with respect to the same datum. It will be evident that if a level surface or line is established and the vertical distances from all the points to the line are measured, a little simple arithmetic will enable the desired heights or reduced levels to be calculated.

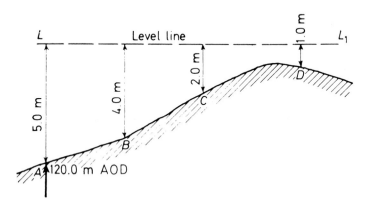

Figure 8.3

In *Figure 8.3*, L–L1 is a level line, A is a point at 120 m AOD, and the heights of B, C and D above OD are required. The vertical distances from A, B, C and D to the level line are measured using a suitable graduated rod. The distances are, respectively, 5.0 m, 4.0 m, 2.0 m and 1.0 m.

The reduced level of B is

$$120.0 + 5.0 - 4.0 = 121 \text{ m}$$

the reduced level of C is

$$120.0 + 5.0 - 2.0 = 123.00 \text{ m} \quad \text{or} \quad 121.0 + 4.0 - 2.0 = 123.0 \text{ m}$$

and the reduced level of D is

$$120.0 + 5.0 - 1.0 = 124.0 \text{ m} \quad \text{or} \quad 123.0 + 2.0 - 1.0 = 124.0 \text{ m}$$

Note that the simple calculation may be done in either of two ways – the significance of these two procedures will be considered later.

In practice, it is not possible to establish a level line, but it is practicable to set up a horizontal line or plane through a point. Since at normal range (under 100 m) the horizontal and level lines through a point are indistinguishable, a horizontal plane or line is set up and the vertical distances are measured from the ground points to the line with a graduated rod. (See also Section 1.6.2.)

8.4.1   Methods of obtaining a horizontal plane or line

The force of gravity may be used in several simple mechanisms to define a horizontal line. If a horizontal line is swung round about a point it will, of course, trace out a horizontal plane. A weighted pendulum, freely suspended, defines the direction of gravity. If a cross-piece is fixed accurately at right angles to the pendulum, a horizontal line is defined by sighting along the cross-piece. The plumb-bob and line has always been used in this way, but the device is clumsy and inaccurate. The pendulum principle, however, has been made use of in a number of modern instruments. If a U-tube is part-filled with water, the water surfaces in the two vertical arms will be level and a horizontal sight line is obtained by sighting over the tops of the two surfaces. This system was used by the Romans and into the seventeenth and eighteenth centuries, but the disadvantages for survey applications are evident.

As in theodolites, a simple, yet most effective, device for defining a horizontal line is the *spirit-level vial*, first developed in 1666. In the early versions, a glass tube was bent to a slight curve, the tube then part-filled with fluid and re-sealed. The bubble formed in the tube seeks the highest part of the tube and when the bubble is centred in the length of the tube the longitudinal axis of the bubble is horizontal. This may be seen in any mason's level on a building site. In modern versions the tube is actually ground to a barrel-style shape rather than being a simple bent tube.

The spirit bubble tube may be attached to a straight-edge, allowing levels to be transferred over a metre or so. The distance can be increased by fixing sights to the straightedge, but the naked-eye range is very limited, particularly for reading graduations on a staff. The obvious way to increase the sighting range was to attach the tube to a telescope rather than a straightedge, and provide a *graticule* within the telescope to act as a 'sight' for aiming the line of sight or *collimation*. The telescope, complete with a graticule in the focal plane of the eyepiece, has a spirit-level vial attached to it in such a manner that the *collimation* (sight) line and the bubble tube axis are parallel. The whole arrangement is supported on a tripod for stability, with provision for the device to be tilted until the spirit bubble is central in the length of its tube, thus making the bubble axis horizontal and therefore the collimation line horizontal. The markings on a graduated staff are observed through the telescope and give the vertical distance from the ground point to the horizontal collimation line.

Automatic levels, developed and produced in the early 1950s, do not rely on a spirit level. Instead, these typically use an arrangement of reflecting prisms fitted within the telescope barrel. The prisms are arranged in the line of sight from the surveyor's eye to the distant staff, and if the telescope is tilted the prism arrangement adjusts its position automatically to compensate for the deviation from the horizontal. The design of the compensating unit or pendulum compensator differs between the various models produced.

Other modern levels developed over the last few years generate a beam of light which may be rotated in a horizontal plane. The beam may be invisible infra-red, or a visible helium neon beam. These instruments, generally termed *lasers*, use either the pendulum principle or the spirit-level to obtain a horizontal plane or line. Lasers, referred to in Chapter 17, are mainly used in setting out works.

8.5   Levelling equipment

The basic equipment for levelling includes a surveyor's level and a levelling staff. Manufacturers produce a wide range of these, many very similar to one another. The items commonly in use are described in the following sections.

8.5.1   The level

Two types of level are in common use today, the *tilting level* and the *automatic level*. Occasionally a *dumpy level* may be encountered, but these are generally considered to be obsolescent now. Many building site operatives refer to any small surveyor's level as a dumpy, generally incorrectly.

8.5.1.1   The dumpy level

The principal features of the dumpy are illustrated in *Figure 8.4*. The essential feature is that the telescope (with spirit-level vial attached) is rigidly fixed to a vertical axis spindle which rotates within and above the levelling head. The *levelling head* is similar to that of a theodolite.

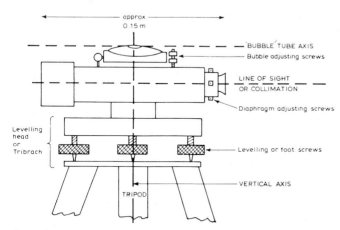

Figure 8.4

With this construction, the telescope can be turned in any direction in the horizontal plane and the vertical axis (and hence the telescope and bubble tube) may be tilted in any direction by appropriate rotation of the foot screws. The bubble tube must be attached with its axis at right angles to the vertical axis. If the foot screws are manipulated until the vertical axis is truly vertical, the bubble axis will be horizontal. If the collimation line is parallel to the bubble tube axis, then rotation of the instrument about the vertical axis will result in the collimation line sweeping out a horizontal plane. The action of making the vertical axis truly vertical is termed levelling up, and is carried out by using the foot screws to centre the bubble when the telescope is alternately placed in two directions at right angles to one another in plan.

It will be evident that two conditions are critical: (1) bubble tube axis at right angles to vertical axis and (2) collimation line parallel to the bubble tube axis. If either of these conditions is not fulfilled, accurate work is impossible. The conditions are difficult to maintain and must be checked and corrected at intervals. Such corrections are termed *Permanent Adjustments* – when once made they should remain correct for months or longer, depending on how the instrument is used.

8.5.1.2   The tilting level

This instrument, first produced about sixty years ago, is a more advanced design than the dumpy and it is generally quicker in use due to the shorter time required to set it up and prepare it for use. The principal constructional features are shown in *Figure 8.5*. The telescope, bubble tube and diaphragm ('cross-hairs') are similar to those of the dumpy but the levelling head may be of either the ball-and-socket or the three foot screw type.

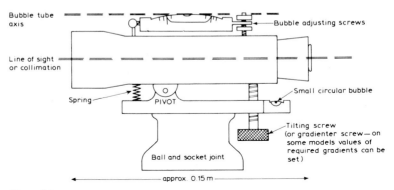

Figure 8.5

Unlike the dumpy, the telescope is not rigidly fixed to the levelling head, but instead is supported on a pivot which allows the telescope to be tilted at an angle to the levelling head – hence the name *tilting*. As with the dumpy, the bubble axis and the collimation line should always be parallel. In use, the instrument is levelled-up roughly by the foot screws or the ball-and-socket, as judged by a small circular spirit-level and is levelled exactly by the *tilting screw* immediately before making the observation on the distant staff.

This arrangement obviates the tedious levelling-up needed with the dumpy, the rough levelling takes very little time, but it is essential that the bubble be centred by the tilting screw before every observation.

The tilting of the telescope at every sight should have no effect on the height of the collimation line and the centring of the bubble takes only a second or two. It will be evident that this instrument has only one critical condition – the collimation line and the bubble tube axis must be parallel. The tilting level therefore has only one permanent adjustment as against the two for the dumpy. The actual adjustment, in practice, is made by moving the bubble tube rather than the diaphragm and modern instruments may have no diaphragm adjusting screws. The tilting level gained its popularity over the dumpy primarily due to the speed of setting up and operation. As well as the features shown in *Figures 8.4* and *8.5*, both dumpy and tilting levels have a means of focusing the 'cross-hairs' and the object being viewed. In addition, many levels have a clamp and slow-motion screw for locking and moving the telescope slowly when it is set up in a horizontal plane. Some levels also have a simple *horizontal circle* which may be useful on occasion when setting out and a small circular bubble for approximate levelling-up of the instrument.

The tilting screw may be graduated in such a way that gradients can be set out and it is then termed a *gradienter screw*. Levels for better accuracy work are generally equipped with a *coincidence prism bubble reading system*. In some instruments, the telescope may be rotated 180° about its longitudinal axis to allow two readings of the staff and instrumental errors are eliminated by taking the mean value. These are sometimes known as *reversible levels*.

8.5.1.3   The automatic level

The essential features of this instrument are shown in *Figure 8.6*. The telescope is rigidly fixed to the vertical axis, as in the dumpy level and the levelling head may be either ball-and-socket or three-foot screw type. A small circular bubble is fitted for levelling the instrument up approximately after it has been attached to the tripod. Once the instrument has been

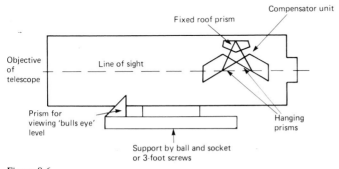

Figure 8.6

approximately levelled-up, the compensator unit operates automatically, its movement being slowed down by damping mechanisms.

When the telescope is aimed in the required direction, a horizontal ray of light entering the centre of the objective lens is passed through a system of fixed and suspended prisms and is directed by these to the centre of the cross-hairs in the diaphragm, where it is observed through the usual eyepiece. Individual manufacturer's arrangement of the mechanisms vary, but in general, provided the telescope is levelled up initially within ± 10 min of arc of the horizontal (as can be achieved using the circular bubble), then in whatever direction the telescope is turned, the cross-hairs will sweep out a horizontal plane of constant height. Despite the damping device, vibrations due to blustery weather conditions or perhaps occasionally traffic or site plant, may make sighting difficult. This vibration may be restricted by laying a hand lightly on the tripod, but it must be remembered that this should never be done with other types of level, since it would disturb their level bubble settings. Some early instruments had problems with excessive friction at the pivots of the suspended prisms, but these appear to have been overcome now.

Fieldwork may be carried out about twice as fast with the automatic level as compared with the other types and fewer mistakes occur because there is no bubble to be continually checked and adjusted. In addition, automatic levels give an erect (right-way-up) image, while the majority of dumpy and tilting levels give an inverted image. Automatic levels are more expensive than the other types, but their advantages are such that they may be expected to replace the others completely.

There is usually only one permanent adjustment to an automatic level, to ensure that the collimation line is horizontal when the instrument is set up. The mode of adjustment depends on the manufacturer – in some cases the diaphragm must be moved, in others an adjustment must be made to the actual compensator unit itself.

In addition to levels being known as dumpy, tilting or automatic, they may be classified according to a theoretically obtainable accuracy in a double run of levels over a specified distance, typically one kilometre. The three main classes are Class I, Class II, and Class III, otherwise described as *precise or geodetic levels, general purpose or engineer's levels* and *construction or builders' levels*. (Hence descriptions such as 'Builder's Automatic', 'Engineer's Tilting Level', etc.)

Where a manufacturer quotes an accuracy figure of 1 mm or less per double run of levels over a kilometre, then it may be considered to be a precise or Class I level. Between 1 and 5 mm would indicate an engineer's level, and 5 mm or more a builder's level. On site, a good set of unadjusted field readings may have a miclosure of up to 2.5 or 3 times the manufacturer's quoted figures.

8.5.2   The staff

As explained earlier, the vertical distance from the ground point to the collimation line is measured with a graduated rod. This rod is termed a *levelling staff* or *levelling rod*. Ordinary staves are made of wood or of aluminium alloy, precise staves are made of a strip of invar steel supported by a wooden or metal frame.

Staves are usually between 2 and 5 m in length, the construction being telescopic, or rigid one-piece, or hinged, or even of jointed sections. The most popular type in the UK is probably the wooden telescopic, which is readily put into use. Since the timber swells when wet, such staves should not be placed in water as the sections will stick and can be neither closed nor extended. These staves are fitted with a brass spring catch to keep the extended section in position and this catch must be checked to ensure that it 'clicks home' and locks the sections properly.

Aluminium staves, less popular, are not unduly heavy, they are resistant to water and are strong and durable. A variety of graduation patterns are available, the commonest being the 'E' pattern specified by BS 4484: Part 1: 1969 and the 'E-and-checkerboard' patterns popular in Europe, both shown in *Figure 8.7*. These types of staff have 10 mm graduations, read by estimation to 1, 2, or 5 mm, generally figured in black on white, although sometimes alternate metres are in red on white. Some users prefer staves with a yellow background, as this can provide a better contrast than white. Reflective facings are available, for night work with an electric torch attached to the level telescope.

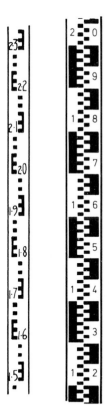

Figure 8.7

These relatively coarse graduations are used because of the wide range of distances over which they can be read in ordinary levelling. In precise levelling, the sighting distances are short and it is usual to graduate staves in the same way as scales, as shown in *Figure 8.8*. Note that the precise staff has two sets of graduations, offset from one another, to enable a check observation to be made. Precise staves with a single set of graduations are obtainable.

A variety of other types of staff have been made, the most notable probably being the *Philadelphia pattern target staff*. This is fitted with a target and vernier scale, and the observer directs the staff man to move the target up or down on the staff until it is on the collimation line. The staff man then reads the staff and obtains the final place of decimals from the vernier scale. Popular in the USA, this is little used in the UK. This form of staff is advantageous for wide river crossings where the staff graduations cannot be read with any great accuracy, but simple home-made targets on a one-piece rigid staff have been found suitable for this.

Figure 8.8

Normal levelling staves can be obtained with the numbers inverted – the figures appear the right way up when observed with a level which gives an inverted image. These staves are a handicap, however, when used with modern erect image automatic levels.

8.5.3   Levelling staff accessories

8.5.3.1   *The staff support or change plate*

This is a triangular steel plate with a raised centre and the three corners turned down, used to support the staff on soft ground and prevent it sinking. Also known as a *shoe* or a *crow's foot*, it is generally fitted with a length of chain for carrying and for pulling it free from the ground (see *Figure 8.9*). A large, round-topped stone provides a suitable alternative.

8.5.3.2   *The staff bubble*

This is a circular spirit level, used to check the verticality of the staff when making observations. It is attached to the back or side of the staff.

8.5.3.3   *The hanging bracket*

This is used with an O.S. bronze flush bracket BM, as shown in *Figure 8.2(b)*.

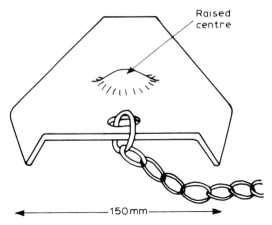

Raised
centre

150 mm

Figure 8.9

8.5.3.4   *Handles and steadying rods*

These are usually supplied with precise staves. Handles are sometimes permanently attached, or may be clamped on to all types of staff. Steadying rods are normally used only in precise work, the staff holder holding a handle and a steadying rod with each hand so that a tripod-like structure is formed, with the staff vertical. In ordinary work, in windy conditions, it may be helpful to use ranging rods as steadying rods.

8.5.4   The parallel plate micrometer

Even with a good telescope and a staff marked in fine divisions, staff readings cannot be made finely enough by simple telescope for precision levelling demanding accuracies such as 0.5 mm per km or so. The *plane parallel plate micrometer* is an attachment to a level which typically permits the determination of level staff readings to 0.1 mm directly and by estimation to 0.01 mm (0.000 01 m).

The device is simply a piece of glass with parallel plane faces, placed in front of the telescope objective and supported on horizontal pivots with the plane faces at right angles to the collimation line. Since glass refracts a ray of light entering it, rotation of the parallel plate causes the collimation line to be raised or lowered while still remaining parallel to its original path. The physical constants of the glass being known, the vertical displacement of the collimation line can be calculated for a known tilt of the plate. The plate is tilted by a *micrometer screw* which registers the *displacement* of the collimation line rather than the *amount of tilt*.

The simplest version, often used as an attachment to a level, has a displacement scale engraved on the edge of the micrometer screw operating the plate. When the device is permanently 'built-in' to the level, the plate is generally linked up to an *optical scale* viewed in an eyepiece alongside the telescope eyepiece.

Figure 8.10 shows the system used on a precise level. The total vertical displacement possible is 10 mm, and the eyepiece scale is graduated 0, 1, 2, . . . 10, each number representing 1 mm of vertical displacement. Each

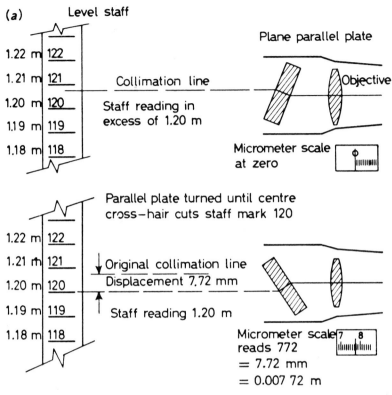

Figure 8.10   Final reading 1.207 72 m

division is further sub-divided into ten parts of 0.1 mm, and these may be sub-divided by eye to 0.01 mm. Operation is extremely simple – after carefully focusing and levelling the instrument, turn the micrometer screw until the central horizontal cross-hair cuts a 10 mm mark on the observed staff, note that reading and add on the reading from the micrometer scale. On drum instruments, take the reading from the edge of the micrometer drum.

8.6   Levelling fieldwork

All levelling operations consist of observing and recording height readings at two or more staff positions from each instrument station. The work may involve only one instrument station, or there may be several instrument stations involved. Each instrument station with its associated set of staff readings may be completely independent of any other set of readings from another instrument station, but more often separate sets are linked by observations on a staff position common to both sets. Examples of these situations are illustrated in *Figures 8.11* and *8.12*. *Figure 8.11(a)* illustrates what is sometimes known as one set-up levelling and the other figures as series levelling or lines of levels. A line of levels run simply to check the accuracy of the work is also known as flying levels.

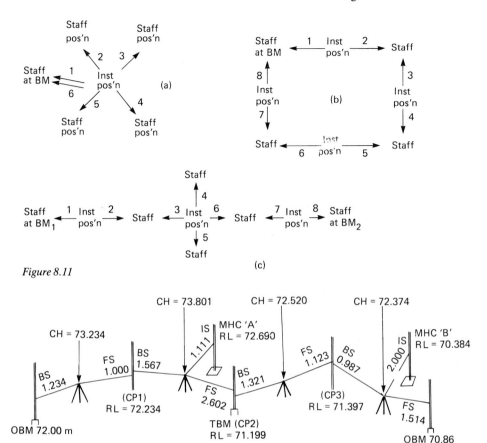

Figure 8.11

Figure 8.12

Abbreviations:

BS — Backsight	FS -- Foresight	OBM — Ordnance bench mark
CH — Collimation height	IS — Intermediate sight	RL — Reduced level
CP — Change point	MHC — Manhole cover	TBM — Temporary bench mark

Each levelling operation, e.g. establishing TBMs, heighting along the line of a section, fixing height pegs on a construction site, checking a particular gradient, etc., may differ in the actual detail of distances to BMs, size of job, labour and equipment available, and so on, but in each case the principles of the operations are essentially the same.

8.6.1 Terms used in levelling

Some levelling terms were defined in preceding sections of this chapter. The following should now be noted.

Backsight: the first reading taken from any instrument station.
Foresight: the last reading taken from any instrument position.

Intermediate sights: readings which are neither the first nor the last to be taken from an instrument position.

Changepoint (CP): a staff position on which first a foresight reading from one instrument position and then a backsight reading from another instrument position are taken.

Collimation height: the calculated height of the line of collimation above or below the datum surface.

Rise and fall: the vertical distance between two consecutive staff positions is either a rise or a fall, a *rise* being a positive difference (the second point being higher than the first) and a *fall* being a negative difference (the second point lower than the first).

These last two terms are also used to identify the two methods commonly used for calculating reduced levels, as will be seen in Section 8.6.6.

8.6.2   The location of staff and instrument positions

The location of staff and instrument positions will vary with the task to be carried out, the equipment in use and the climatic and environmental conditions. However, the instrument and staff must always be set up on firm ground if this is at all possible and the task should be carried out with as few instrument stations as practicable, within the limitations on observing imposed by the length of the levelling staff and the horizontal distance. Distances between instrument and staff should be kept uniform, as far as possible, especially when levelling over long distances, or levelling up or down steep gradients, or when establishing temporary bench marks.

The purpose of equalizing these distances is to minimize instrument errors and also to reduce the effect of the Earth's *curvature* and the *refraction* ('bending') of light by the Earth's atmosphere. In most surveying tasks the effects of curvature and refraction are generally so small that they may be ignored for practical purposes. The combined error due to refraction and curvature in levelling is approximately 24 mm at a distance of 600 m and 0.25 mm at 60 m distance.

The ideal length of sight between instrument and staff is from 45 to 60 m, longer sights tending to lead to inaccuracies in reading and shorter sights implying more instrument stations and a more costly job in terms of time and money. The length of the levelling staff will impose restrictions on the length of sight on steep slopes, especially if equal sight lengths are to be maintained. Further, it is best to avoid reading the lower 0.2 m of the staff since refraction has its greatest effect near the ground, particularly in precise work. *Grazing rays* (sight line skimming the ground) should be avoided when sighting over the crest of a hill. To maintain equal sights, it may be necessary to level up or down hill in a zig-zag pattern in plan, or alternatively to select the high spots, if any, as instrument stations.

8.6.3   Setting up the level and tripod

The level and tripod are set up in a manner similar to the theodolite and, again, the operations are known as the temporary or station adjustments.

8.6.3.1   Setting up the tripod

This is as described for the theodolite in Chapter 5, except that no attempt is made to set the tripod exactly over a ground mark since this is not required in levelling.

8.6.3.2   Attaching the instrument to the tripod

This is again similar to the theodolite, there generally being some form of captive bolt in the tripod head, screwed up into the underside of the level. On older levels, the top of the tripod may be threaded and the whole instrument screwed on to the tripod head.

8.6.3.3   Levelling-up the instrument

The dumpy: the procedure is the same as levelling-up the horizontal or upper plate bubble of a theodolite (see Chapter 5).

The tilting and automatic levels: with these instruments, the levelling-up is a much quicker and easier process, since it is only necessary to get the small circular bubble approximately central.

If the levelling head is of the three-foot screw type, then the foot screws should be used as needed to centre the bubble, using a technique similar to that used for the theodolite. In this case, however, there is no need to turn the instrument in plan.

Where a ball-and-socket levelling head is fitted, the clamping ring or fastening screw, as appropriate, should be eased off with one hand while the other holds the telescope and tilts it as necessary until the circular bubble is central. Finally, the ball-and-socket should be clamped without disturbing the centring of the bubble.

In the case of an automatic level, if the instrument is in adjustment the compensator unit should now ensure that the line of collimation is horizontal. With the tilting level, the tilting screw must be used to centre the main bubble immediately before taking each reading.

When using an automatic level for precise work, it may be sensible to assume that the compensator unit has some residual error causing the line of sight to be inclined to the horizontal. This can then be considered to be a systematic error, eliminated by an appropriate observing technique, provided that the small circular bubble is always levelled in the same manner; for example, the final adjustment of the circular bubble always being made towards No. 1 staff. (In precise work, the use of two levelling staves is recommended.)

8.6.3.4   Focusing and the elimination of parallax

These are as described for the theodolite.

8.6.4   Observing the staff

The procedure to be used depends upon the type of instrument in use, as shown in the following sections.

8.6.4.1   Dumpy level

Point the telescope at the staff by aiming over and along the telescope barrel (better instruments are fitted with open sights). Focus carefully on the staff. Turn the telescope until the vertical cross-hair bisects the middle of the staff. Some levels have a clamp and slow-motion screw for 'fine-pointing'. Glance at the bubble, ensure that it is still central or in its mean position. If not, re-centre the bubble by the appropriate foot screw. Read and record the staff graduation at the central horizontal cross-hair. After noting, glance at the staff again as a check.

8.6.4.2   Tilting level

Aim, focus and bisect the staff as described for the dumpy. Centre the longitudinal bubble using the tilting screw. Levels without a coincidence prism bubble reading system generally have a mirror (at 45° above the bubble tube) in which the bubble may be viewed. The mirror should be used for all bubble viewing, since its actual position and its apparent position as viewed in the mirror are not identical. Read and record the staff graduation at the centre horizontal cross-hair.

8.6.4.3   Automatic level

Aim, focus, bisect the staff as above. While looking through the telescope, tap the tripod lightly to ensure that the prism system is operating freely – the horizontal centre hair should move slightly but stabilize quickly. Read and record the staff graduation.

8.6.4.4   Telescope reticule and stadia measurement

Most modern levels have a *reticule* similar to that of a theodolite, the upper and lower horizontal 'hairs' being termed *stadia lines*. The stadia hairs may be used to determine the distance between the instrument and the staff, correct to within 0.2 m over the recommended maximum sighting distance of 60 m, an accuracy more than adequate for maintaining equal sight lengths. The instrument makers have positioned the horizontal lines of the reticule so that they are equally spaced and in such a manner that if the staff is read against all three hairs, then the horizontal distance from instrument to staff is usually equal to 100 times the staff intercept between the upper and lower hair readings. (*Figure 8.13.*)

It should be noted that, except in precise levelling, the sighting distances are not checked in this way at every station, rather it is customary to rely on the staff holder to ensure equal distances by pacing. The stadia hairs may also be used, particularly for an uncertain beginner, to provide a check on the centre hair reading, since the average of the upper and lower hair readings should be equal to the centre hair reading. In precise work it is usual to observe as follows:

No. 1 staff, stadia hairs, centre hair
No. 2 staff, centre hair, stadia hairs
No. 2 staff, centre hair
No. 1 staff, centre hair.

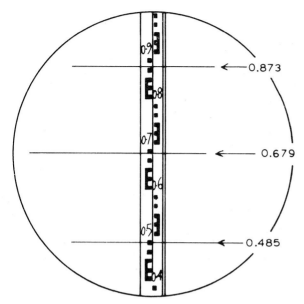

1 Has the centre hair been read correctly ?

    0.873
  + 0.485
    ‾‾‾‾‾
    1.358 ÷ 2 = 0.679
                    = centre hair

2 What is the horizontal distance ?

    100 x (0.873 - 0.485)
  = 100 x 0.388
  = 38.8 m

Figure 8.13

In all cases, of course, both the staff and the parallel plate are read and recorded. This technique gives a check on both the distances and the centre hair readings. It will be noted that the centre hair readings on both occasions are read consecutively to minimize errors resulting from the staves or the instrument sinking or rising.

8.6.5 Duties of the staff holder

A good staff holder is essential if the surveyor is to carry out the levelling accurately and quickly and such an assistant will often anticipate the surveyor's requirements. Where possible, the surveyor should advise the staff holder, before the commencement of the levelling, as to where staff positions will be required. The staff holder should

  ensure that the staff and ancillary equipment are in good working order;
  lay the staff on a clean, dry area of ground when it is not in use;
  carry the staff between stations, holding it vertically and retracted if it is a telescopic type;

check before setting up the staff that its zero end and the raised centre of the staff support are free from dirt;

select a staff position which may be easily re-located later, if requested to do so;

use a staff support, or a large stone, if the ground is not as firm as the surveyor would wish for a staff position (a staff support should always be used in precise work);

extend or clamp the staff as necessary, ensuring that this has been done correctly;

understand the simple system of hand signals used for communication in the field (similar to chain survey);

place the zero end of the staff gently on to the selected staff position, as required;

hold the staff vertically for readings, keeping the hands off its face so as to avoid dirtying it or obstructing vision;

check that the staff remains vertical during readings by occasionally glancing at the staff bubble;

move the staff or its support only when instructed to do so;

inform the surveyor if the staff (or its support) is displaced or tending to sink (possibly under its own weight, or through the staff holder inadvertently leaning his weight on it);

maintain equal distances (within about a metre) between an instrument and its foresight and backsight staff positions by careful pacing:

wipe clean and dry off the staff and ancillary equipment upon completion of the day's levelling.

8.6.6   Level booking, recording and reducing field observations

For the majority of levelling tasks readings are recorded in a ruled level book, the form of ruling depending on the surveyor's preference. The standard patterns are *Rise and Fall*, and *Collimation Height*. In precise work a modification of the Rise and Fall form is commonly used so as to enable the stadia hair and parallel plate micrometer readings to be booked. Because the recording and reduction of levelling observations is a relatively simple task, it is possible to program the more sophisticated pocket calculators, or hand-held computers, as data loggers, also known as electronic field books and data acquisition devices. A good program should prompt the user as to the data to be entered and, at completion, determine the misclosure and display or print out the reduced levels.

8.6.6.1   The rise and fall method

The line of levels shown diagrammatically in *Figure 8.12* is used to illustrate this method in *Figure 8.14*. The following points should be noted in *Figure 8.14*.

Each line of the book represents a staff position, and that position is identified in the 'Remarks' column. The first reduced level is entered from the given data, the remainder being calculated as follows:

Obtain the rises or falls, thus
1.234 − 1.000 = + 0.234 = rise
1.567 − 1.111 = + 0.456 = rise
1.111 − 2.602 = − 1.491 = fall
1.321 − 1.123 = + 0.198 = rise
0.987 − 2.000 = − 1.013 = fall
2.000 − 1.514 = + 0.486 = rise

Note carefully where these calculated values have been entered on the booking sheet, also note that while each backsight or foresight is used only once, each intermediate sight value is used twice.

Date _____ Levels taken for _____

From _____ To _____

BACK SIGHT	INTER-MEDIATE	FORE SIGHT	RISE	FALL	REDUCED LEVEL	REMARKS
1.234					72.000	OBM St Johns Church
1.567		1.000	0.234		72.234	CP1
	1.111		0.456		72.690	MHC 'A'
1.321		2.602		1.491	71.199	TBM (CP2)
0.987		1.123	0.198		71.397	CP3
	2.000			1.013	70.384	MHC 'B'
		1.514	0.486		70.870	OBM (70.86) The Ring of Bells 'PH'
5.109		6.239	1.374	2.504	72.000	
6.239			2.504		−1.130	
−1.130	✓		−1.130	✓		

Figure 8.14

Using the calculated rises and falls, calculate the reduced levels in succession from:

Reduced level = reduced level on previous line + rise between the two staff positions, *or* − fall between the two staff positions.
e.g. 72.000 + 0.234 = 72.234
     72.234 + 0. 456
     = 72.690
     72.690 − 1.491
     = 71. 199
     71.199 + 0.198 = 71.397, etc.

The calculations may be summarized as:

(BS1 or IS1) − (FS2 or IS2) is a rise if positive *or* a fall if negative
RL1 + rise2 *or* − fall2 = RL2.

The commonest mistakes which occur in level booking and reduction are arithmetical, hence every arithmetic operation must be checked. If the calculations above are correct, then:

The sum of the backsights − the sum of the foresights = the sum of the rises − the sum of the falls = the last reduced level − the first reduced level.

Note that in practice, the correct routine is to calculate the rises and falls, then compare the difference between the backsight and foresight sums with the difference between the sum of the rises and the sum of the falls, to ensure that the rises and falls have been accurately computed. Only when this arithmetic has been checked should the reduced levels be calculated. Finally, the checking should be completed by comparing these differences with the difference between the first and last reduced levels. The Rise and Fall method provides a complete check on the arithmetic of the reductions, but it must be appreciated that it does not check the accuracy of the actual observations. These may be checked or 'proved' only by levelling back to the opening bench mark or completing the line of levels on to another point of known height and comparing the calculated and known heights.

It will be noted that in the above example there is a *misclosure* of 10 mm, since the calculated reduced level at the end of the line of levels is 70.870 m, while the given height is 70.86 m. The method of dealing with this is described in Section 8.6.6.6.

8.6.6.2   The collimation height or height of instrument method

In this method, the level bookings are exactly the same as for the rise and fall method, the difference lying in the method of reduction. Again, the line of levels in *Figure 8.12* is used in *Figure 8.15* to illustrate booking and reduction by this method.

BACK	INTER-MEDIATE	FORE.	COLLIM-ATION	REDUCED LEVEL	REMARKS
1·234			73.234	72·000	OBM St Johns Church
1·567		1·000	73·801	72·234	CP1
	1·111			72·690	MHC 'A'
1·321		2·602	72·520	71·199	TBM (CP2)
0·987		1·123	72·384	71·397	CP3
	2·000			70·384	MHC 'B'
		1·514		70·870	OBM (70.86)'The Ring of bells'
5·109		6·239		72·000	
6·239				-1·130	✓
-1·130					
	Check on intermediate sights				
	3·111	6·239	438·124	428·774	
				3·111	
				6·239	
				438·124	

Figure 8.15

Points to note in *Figure 8.15*.

As before, each book line represents a staff position, that position being identified in the 'Remarks' column. The first reduced level is again entered from the given data, but the remainder are calculated as follows:

Reduced level + backsight = *collimation height*, and collimation height − foresight (*or* intermediate sight, as relevant) = reduced level.

Thus

72.000 + 1.234 = 73.234 (inst ht)
73.234 − 1.000 = 72.234 (RL of CP 1)
72.234 + 1.567 = 73.801 (inst ht)
73.801 − 1.111 = 72.690 (MHC 'A')
73.801 − 2.602 = 71.199 (RL of TBM)
71.199 + 1.321 = 72.520 (inst ht)
72.520 − 1.123 = 71.397 (RL of CP 3)
71.397 + 0.987 = 72.384 (inst ht)
72.384 − 2.000 = 70.384 (MHC 'B')
72.384 − 1.514 = 70.870 (RL of OBM)

Again, the position where these values have been entered on the booking sheet should be noted carefully. The calculations could be summarized as

RL + BS = CH
CH1 − FS2 = RL2, or
CH1 − IS3 = RL3

As always, the arithmetic should be checked. The difference between the sum of the foresights and the sum of the backsights should equal the difference between the first and last reduced levels. If this is so, it checks the calculation of the changepoint reduced levels, but it does not check the calculation of the reduced levels of the intermediates.

A method is available for checking the intermediate reductions, the rule being 'The sum of each collimation height multiplied by the number of reduced levels obtained from it is equal to the sum of all the intermediate sights, foresights and reduced levels excluding the first reduced level' (see *Figure 8.15*).

This check is so tedious that it is doubtful if it is used at all in low accuracy work and precise work is reduced by the rise and fall method. The collimation height method is widely used in ordinary building works and since this is often levelling carried out from a single instrument station then collimation height reduction is fast and easy. The commonest errors in levelling, however, are arithmetical, particularly with individuals who seldom use a level. Such persons and beginners, would be best to use the rise and fall method to eliminate such mistakes.

8.6.6.3   *Checking level entries extending over more than one page*

Where level book entries extend over more than one page, each page should be separately checked when reducing, although with the greater use being made of pocket calculators this is perhaps not so important today. If each page is checked and this is strongly recommended, then three or four lines must be left clear at the bottom of each page or booking sheet to allow for totals and differences. Where a page is to be checked separately, then the entries on the page must commence with a backsight and end with a foresight. It will be evident that the last reading on a page will generally be either a reading to a changepoint or a reading to an intermediate sight.

If the last entry is for a change point, then the foresight reading to that point will be the last reading on the page and the backsight reading to the same point must be entered as the first reading on the next page. If the page ends at an intermediate sight, then it must be booked as a foresight and repeated again as a backsight as the first reading on the next page. Note that a page can only be checked by the methods above if there are the same number of backsights and foresights on the page.

8.6.6.4   Using the staff in the inverted position

On occasion it is convenient to use the staff upside down where a point to be levelled is above the collimation line, e.g. bridge soffit levels, or a line of levels run across a high wall. In this case, the inverted staff readings should be booked as negative values and due account of the negative sign taken in the reduction calculation. A sketch may help to clarify the required arithmetic.

8.6.6.5   Permissible error in levelling

The example reductions in *Figures 8.12, 8.14* and *8.15* show a misclosure of 10 mm or 0.010 m. Since the arithmetic has been checked, the 0.010 m must be an error in the levelling and not in the calculations. Since there are always errors in survey, a limit has to be set for the *permissible* (i.e. acceptable) *error* in any levelling job. The actual error permissible depends upon the type of job. For ordinary careful work, on fairly flat ground, a reasonable allowance may be taken as $\pm 20\sqrt{K}$mm, where K is the total distance levelled in kilometres (length of circuit or distance from BM to BM).

Where the same work is carried out on steep slopes, or levelling for earthworks volumes or contours, $\pm 30\sqrt{K}$mm. For more accurate work, with equal backsight and foresight distances and careful estimation of the third decimal place (i.e. reading carefully to 1 mm or 0.001 m), then $\pm 10\sqrt{K}$mm may be reasonable, e.g. in establishing TBMs. By contrast, the typical allowance in precise levelling might be $\pm 2\sqrt{K}$mm. In the survey of small sites, closing errors not exceeding $\pm 5\sqrt{s}$ mm are generally acceptable, where s = the number of instrument set-ups used.

Commonsense must be applied in deciding whether a misclosure is acceptable or otherwise. As an example, if reduced levels were required for contouring only, then a misclosure of $100\sqrt{K}$mm might be adequate, although such a standard should be discouraged in levelling generally. When the permissible error for a task has been exceeded and the error cannot be located in one part of the levelling, it is necessary to repeat the whole of the levelling. For this reason, it is best to carry out a check over the same changepoints as the original levelling – the error may be located in one section.

8.6.6.6   Adjustment of the level book

In ordinary levelling, the misclosure is likely to be a combination of both accidental (random) errors, e.g. small errors in reading the staff and cumulative errors, e.g. the instrument and/or staff sinking during the intervals between taking readings.

Although accidental errors may occur in all readings, such errors in intermediate readings do not affect the overall misclosure, since they are not carried forward and do not influence the reduced levels of the changepoints. If a staff sinks while being held at a changepoint, this will affect the backsight reading to the staff and the intermediates and foresight following.

If an instrument sinks while set up at a point, the longer it is at the point then the greater will be the error in the foresight reading from that position and the errors in the intermediate readings will gradually increase.

Accidental error should be distributed equally at the changepoints, that is to say all backsights should be adjusted by the same amount while all foresights should be adjusted by the same amounts but of opposite sign. Cumulative error, such as that due to sinking, could be adjusted similarly, but in theory the intermediates from a station should be adjusted by varying amounts. In practice, it is the *reduced levels* which are adjusted rather than the staff readings. Generally the same correction is applied at each changepoint and at the closing BM, the sum of these corrections being equal to the magnitude of the misclosure but of opposite sign. Adjustments are made to the nearest millimetre only in ordinary work.

The result of applying the above rules to the line of levels in *Figures 8.12, 8.14* and *8.15* with an error of +0.010 m could be:

Opening BM	Adjustment	Running total
RL at CP1	−0.002	−0.002
RL at CP2	−0.003	−0.005
RL at CP3	−0.002	−0.007
Closing BM	−0.003	−0.010

There are four points at which the adjustment is incremented, and +10 mm misclosure, giving −2.5 mm at each, rounded here to −2, −3, −2 and −3. As the adjustment at a CP is applied, all reduced levels following that point are adjusted by the same amount, the process being repeated through all the reduced levels. Some surveyors may adjust intermediates differentially, on the assumption of cumulative error, but no rule can be laid down for this. In precise levelling there should be no intermediate sights and cumulative error should be kept to a minimum, then this latter problem does not arise.

8.7   Permanent adjustments to the level

As previously stated, there are certain adjustments which must be maintained in good order if an instrument is to provide accurate work.

These are the *permanent adjustments* of the level and the checking and correcting of these varies with the type of instrument. Levels should be checked regularly, but it should be noted as regards the following sections that for ordinary levelling collimation adjustment is not made unless the error exceeds about 1 mm per 15 m of sight distance.

8.7.1 The dumpy level

The requirements for the dumpy are that (1) the bubble tube axis should be perpendicular to the vertical axis, and (2) the collimation line should be parallel to the bubble tube axis.

To check (1), set up the instrument, centre the bubble carefully in two directions in plan. Turn the telescope through 180° in plan. The bubble stays central if it is in adjustment, if it moves off centre it must be adjusted.

To adjust (1), note the amount the bubble has moved off centre. Using the foot screws, move the bubble halfway back to centre. Move the bubble the remainder of the way back to centre by means of the capstan-headed screws fitted at one end of the bubble tube. When complete, check again and repeat adjustment as necessary.

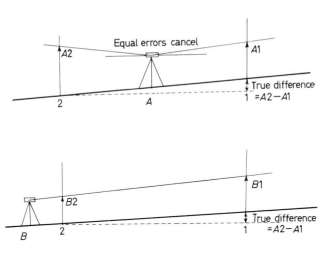

If collimation line at *B* is horizontal,
then *B*2 should equal *B*1 + (*A*2 − *A*1)
Reading *B*1 should equal *B*2 − (*A*2 − *A*1)

Figure 8.16

To check (2), on reasonably level ground set out two stable marks, generally wooden pegs, about 30 m apart (or as recommended by the manufacturer). Set up and level the instrument midway between the pegs, at point A as in *Figure 8.16*. Observe the readings on a level staff held in turn on peg 1 and peg 2, note these as readings A1 and A2. The difference between the readings A1 and A2 will be the true difference in level between the pegs, regardless of any collimation error in the level.

Now set the level up at point B, on the line of the pegs but outside them and as close to point 2 as the short focus distance of the telescope will allow when the staff on 2 is observed. Alternatively, the level could be set up between the marks but positioned so that the eyepiece is only 25 to 50 mm away from the staff held on peg 2, *Figure 8.17*. Note the staff readings on 1 and 2 as B1 and B2. If there is no error, then the difference between B1 and B2 will be the same as that between A1 and A2. If they are not the same there is *collimation error*.

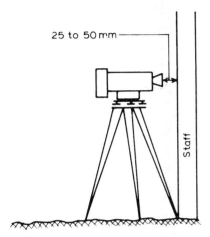

25 to 50 mm

Staff

Figure 8.17

In the alternative method, the reading on to the staff at peg 2 should be taken through the objective end of the telescope. In reading the staff this way, the surveyor may not be able to see the cross-hairs, but this is not important since the field of view will be very small and it is easy to judge the centre of the visible circle. A pencil point laid against the staff face, *Figure 8.18*, will aid in determining the reading, since the pencil will be

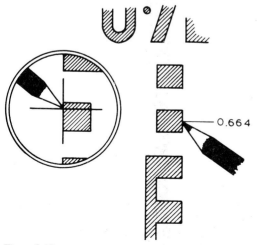

0.664

Figure 8.18

visible through the telescope even if no graduation figures can be seen. The pencil is moved until its point is central then the graduation at the pencil point is observed directly by eye. (For obvious reasons this method of test is described as the *Two-peg test*).

To adjust (2), calculate what the staff reading on B1 should be, assuming there is no error in reading B2. (There may be an error in B2, depending on the method used, but for most practical purposes it can be ignored.) Move the diaphragm in the telescope, using the adjusting screws fitted at the top and bottom of the end of the telescope nearest the eye, until the central horizontal cross-hair cuts the staff on 1 at the computed true reading. Check and repeat if necessary.

8.7.2   The tilting level

This instrument has one requirement only, that the bubble tube axis and collimation line should be parallel (*collimation adjustment*). To check, carry out the two-peg test as described for the dumpy.

To adjust, calculate the required correct reading on staff 1. Using the tilting screw, bring the central horizontal cross-hair to this calculated reading. This will result in the bubble moving off centre. Centre the bubble carefully again by means of its own adjusting screws and the operation is complete. Repeat the check and adjustment again as necessary.

Note the important difference – in the dumpy, the collimation line is moved, by moving the diaphragm, but in the tilting level it is the bubble which is adjusted and not the diaphragm position. Makers' instruction manuals should be consulted for detailed guidance on adjustment of any gradienter screw, for adjustment of precise levels with bubble scales and for reversible levels.

8.7.3   The automatic level

This instrument also has one requirement only, that the collimation line be horizontal when the circular bubble has been centred. To check, carry out the two-peg test as before.

To adjust, calculate the required correct reading on staff 1. Move the collimation line until the correct reading is at the central horizontal cross-hair. The method of moving the line of collimation to get the cross-hairs on to the reading, however, varies between different instruments and reference should be made to the maker's handbook. Some instruments are adjusted by moving the diaphragm, as for the dumpy, while in others it is the compensator unit which must be adjusted. In some, both must be adjusted. It should be noted that in some automatic levels the eyepiece and the objective lens do not lie in the same horizontal plane, hence the method of looking at the staff through the objective lens is not feasible.

8.7.4   Other adjustments

The principal adjustments necessary for the operation of levels have been described above. Other aspects which may require attention include the

TABLE 8.1.

Source	Precaution
Errors attributable to the surveyor:	
Mistakes in reading the staff	Read all three hairs
Mistakes in booking readings	Book stadia readings in 'remarks' column
Disturbing level or tripod or both	Check position of bubble; do not lean on or kick tripod
Failure to level the bubble	Check before reading
Incorrect focusing	Eliminate parallax
Mistakes in reducing levels	Carry out mathematical checks
Errors attributable to the staffholder: (see Section 8.6.5 – Duties of the staff holder)	
Not holding the staff upright	Check staff bubble
At change points, not ensuring that the staff is held in exactly the same position for both back- and foresights at a point	Use staff support, always used in precise work
Unequal back- and foresights	Surveyor check occasionally using stadia hairs, always in precise work
Staff not properly extended	Surveyor check as needed by viewing connecting portion(s) of staff through telescope
Errors attributable to the ground or climatic conditions:	
Sinking/rising of the instrument and/or staff	Set up on firm (not frozen) ground, use staff support
Strong winds, staff and instrument vibrating	Find sheltered spot for instrument and staff, set tripod up low with its legs spread and tread its feet well in, if automatic level, hold tripod lightly; brace staff with two steadying rods; if wind becomes too strong, cease work
Heat shimmer, staff graduations unsteady (appear to bounce)	Reduce length of sights; try to keep line of sight well above ground level
Direct heat of the sun causing differential expansion of instrument parts, bubble tube in particular:	Hold field book to shade bubble tube when levelling up; survey umbrellas can be bought
Curvature and refraction (but no visible disturbance of image)	These errors generally negligible, but reduce or eliminate by using equal sight lengths; keep sights short (max. 60 m); avoid grazing rays (sights near ground), avoid continually reading zero end of staff, e.g. when going uphill. Steep hills should be avoided in precise levelling. In precise work each line is measured twice, at different times on different days Sun near horizon may make sighting impossible. Raindrops on objective may make reading difficult. Avoid sighting near the sun, use telescope rayshade. Use rayshade and umbrella
Errors attributable to the level:	
Faulty permanent adjustments	Check adjustments from time to time (see Section 8.7); keep backsight and foresight lengths equal
Errors attributable to the tripod:	
Play in the joints	Check occasionally; tighten as necessary
Errors attributable to the staff:	
Longitudinal warping of the staff	Errors usually insignificant, but they are cumulative; if serious, discard staff
Graduation errors	An error in the 'zero point' of the staff has no effect, but errors at other points on staff may be cumulative; occasionally check graduations against steel tape
Staff bubble out of adjustment	Check occasionally; error minimal but cumulative

correction of the circular bubble on tilting and automatic levels, the take-up of wear on foot screws, etc. Where an instrument is to be used for any length of time it is advisable to obtain a copy of the maker's instruction manual.

8.8   Sources of error in levelling

As with linear and angular measurement work, levelling is never free from error. Some errors are due to carelessness, some have cumulative or constant or systematic effects on the results of the levelling. *Table 8.1* lists the common sources of error in levelling, together with the precautions to be taken to minimize their effects.

Leicester Polytechnic

LEVELLING RESULTS

Levelling results printout from CBM micro program LEVELS

Copy 1   Dated 7th August 1984 (Data held on file le-levelsdemo)

Levelling survey :   Leicester Royal Car Park Site
                     (Line between two known benchmarks)

Client :   R E Paul

Levelled by   D.Powey - HND Building Year 1 1979/80 on 15th May 1980

Summary of Readings and Reduced Levels

Point No.	Back-sight	Inter sight	Fore-sight	Rise (unadjusted)	Fall	Original Level	Adjusted Level	BM or CP No.
0	1.260					59.230	59.230	Start BM
1	0.010		1.110	0.150		59.380	59.379	1
2	0.313		0.145		0.135	59.245	59.244	2
3	1.502		1.825		1.512	57.733	57.731	3
4		1.960			0.458	57.275	57.273	
5		1.885		0.075		57.350	57.348	
6		1.959			0.074	57.276	57.274	
7		1.543		0.416		57.692	57.690	
8		1.473		0.070		57.762	57.760	
9		1.547			0.074	57.688	57.686	
10		2.268			0.721	56.967	56.965	
11		2.200		0.068		57.035	57.033	
12		2.313			0.113	56.922	56.920	
13		1.788		0.525		57.447	57.445	
14		1.650		0.138		57.585	57.583	
15		1.885			0.235	57.350	57.348	
16	1.738		2.780		0.895	56.455	56.452	4
17		1.565		0.173		56.628	56.625	
18		1.490		0.075		56.703	56.700	
19		1.575			0.085	56.618	56.615	
20		1.093		0.482		57.100	57.097	
21		0.965		0.128		57.228	57.225	
22		1.090			0.125	57.103	57.100	
23		1.775			0.685	56.418	56.415	
24		1.773		0.002		56.420	56.417	
25		1.778			0.005	56.415	56.412	
26		1.650		0.128		56.543	56.540	
27		1.282		0.368		56.911	56.908	
28		1.480			0.198	56.713	56.710	
29		1.778			0.298	56.415	56.412	
30		1.708		0.070		56.485	56.482	

Continued -

Figure 8.19

```
Page 2  Levelling survey :  Leicester Royal Car Park Site
                    (Line between two known benchmarks)

Point   Back-   Inter  Fore-  Rise    Fall   Original Adjusted BM or
 No.    sight   sight  sight  (unadjusted)   Level    Level    CP No.

  31            1.575         0.133          56.618   56.615
  32    1.388          0.871  0.704          57.322   57.318      5
  33    1.358          1.269  0.119          57.441   57.437      6
  34                   0.994  0.364          57.805   57.800  Close BM


        Start    BM    =   59.230

        Closing level  =   57.805
        Closing BM     =   57.800

        Closing error  =    0.005

Total unadjusted fall  =    1.425


    ■MICROSURVEY■ Software by Construction Measurement Systems Ltd
         Copyright (C) W.S.Whyte 1980/1/2/3  Tel: (0537 58) 283
         8000/2.1/8050/28 - for Leicester Polytechnic - 12.1.84
```

Figure 8.20

8.9   Microcomputer applications

The MICROSURVEY program Levels may be used to reduce and adjust observations in ordinary levelling. The field readings are entered in exactly the same order as they were observed, backsight and foresight reading numeric values being distinguished by 'b' and 'f' prefaces respectively. The user must indicate initially whether the levelling is

open,
a closed circuit, or
closed between two bench marks,

and the program will automatically adjust closed levels.

Figures 8.19 and *8.20* show an example printout, using the standard format used in rise and fall reduction, but with both the original and adjusted levels shown and the misclosure. Levels data may be stored on disk and files may be read in and data amended in the event of re-levelling, the standard facilities. A maximum of 201 ground points may be levelled, the ground points (not the readings) being numbered 0, 1, 2, 3, ... Should it be necessary to handle more than this number of points in a task, the changepoints may be abstracted from the level book and reduced by the program, then intermediates reduced separately and fitted to the changepoints.

Levelling is particularly suited to direct field entry with a hand-held micro.

Chapter 9

Applications of levelling

9.1 Introduction

Levelling is used to establish the heights of one or more points, located either in a line over the Earth's surface, or in a series of lines, or covering an area. This chapter deals with the common ordinary levelling tasks of (1) levelling to establish a TBM, (2) levelling for contours, and (3) levelling for sections and cross-sections, together with some consideration of precise levelling and river crossing problems. Levelling for setting out is considered in Chapter 17.

9.2 Establishing a TBM

This is done by running a line of levels (with no intermediate sights) from a bench mark to the proposed TBM, then from the TBM to another bench mark, or alternatively, back to the original bench mark. This procedure allows a check on the calculated level of the TBM.

9.2.1 Operations before starting to level

Before starting to level, obtain the necessary bench mark information, locate the bench marks, select and establish a mark for the TBM. Decide on the route the levelling is to follow and (if considered necessary) check the accuracy of the level to be used.

9.2.2 Levelling the TBM

Set up the level, sight on to the opening (starting) bench mark, read the staff to the nearest 0.001 m, book the reading as a backsight, then check it again. Signal the staff holder to move to the next staff position. (It is important that the surveyor does not leave the level unattended at any time, particularly where there is considerable pedestrian or vehicular traffic, thus the use of hand signals is important.) Sight the staff at the new position, read the staff, book the reading as a foresight, check as before.

Signal the staff holder to remain in position and move the instrument to the next position. Note that at all times either the staff should be in position or the level should be in position – they should never both be off the ground and moving at the same time. When moving the instrument, it may be carried on its tripod with the legs closed and held in a vertical position, but if being transported in a vehicle then the level should be placed in its case and the case held on the lap of one of the passengers.

Set up the level at the new position, sight the staff and take a backsight reading as before, signal the staff holder to move on, then take a foresight reading on the staff and repeat these operations until the TBM is reached. The line of levels must close (finish) with a foresight reading on to the new TBM.

When the TBM has been reached, move the instrument to a new position and repeat the whole process to close a new line of levels on to the second bench mark, or alternatively, back on to the opening BM.

9.2.3   Reducing the levels

The levels should be reduced by the rise and fall method, as described in Section 8.6, and the line should close within $\pm 10\sqrt{K}$mm, where K is the total distance levelled in kilometres.

In establishing a TBM the positions selected for the instrument and the staff should be firm and solid, keeping cumulative errors at a minimum and any misclosure should then be distributed equally through the change-points.

9.3   Contouring plans by level and staff

A *contour* may be defined as a line on a plan or map representing an imaginary line on the ground connecting all adjacent points of equal height. (The best visual contour line is the water's edge of a still lake.) The difference in height between successive contour lines is termed the *contour vertical interval* (VI) and is generally constant throughout one drawing. The shortest horizontal distance between any two contour lines varies with the slope of the ground and is termed the *horizontal equivalent* (HE). The salient features of the ground are readily observed from a study of the contour lines, close contours showing steep slopes, widely separated contours showing gentle slopes and evenly spaced contours a constant slope. Contour lines never cross one another, but several contour lines coming together indicate a cliff or overhang and contours always close on themselves, although they may not do so on the particular map or plan.

Individuals who make infrequent use of contoured maps or plans sometimes have difficulty in deciding whether a particular feature is a ridge or a dry valley, since both appear similar on the drawing, see *Figure 9.1*. The simplest approach is to imagine oneself standing on the ridge or in the valley and looking along the feature, then turn through 90°. An increase in the contour values indicates that one is looking up a hill (in a valley), while a decrease indicates looking down a hill, standing on a ridge.

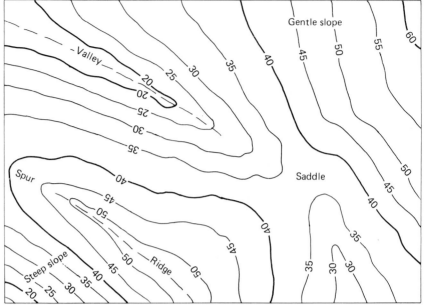

Figure 9.1

9.3.1   Uses of contoured maps and plans

A contoured map or plan should give the user a picture of the ground which will enable the relief of the ground to be interpreted. A ground profile, i.e. a section, could be produced along any required line on the plan, thus determining the location of a route to a suitable gradient, or a check on whether distant points are intervisible, or the suitability of an area for a reservoir, or the extent and volume of earthworks. Contoured plans are often useful for the planning of the layout of estates, although on small building sites, etc., contours are not often demanded. For minor road and drain layout, a few spot heights scattered over the area concerned is often sufficient.

9.3.2   Choice of contour interval

This depends primarily upon the scale of the map and the ground relief, together with the purpose of the survey, although time and the finance available will have an influence. On small building sites a VI of 0.2 to 0.5 m is often recommended, but the plans of such sites are often to very large scales. Large estates, housing or industrial, and reservoirs, then a VI of 0.5 to 2.5 m is frequently recommended. For small-scale plans for roads and railways, often 1 to 5 m is used.

As a guide, *Table 9.1* is recommended (note the similarity of the digits in the centre row). On very steep ground the VI can be greater, and conversely smaller on very flat ground.

TABLE 9.1.

Ground relief	Common plan scales			
	1:200	1:500	1:1000/1:1250	1:2500
Very flat	0.2 m	0.2 m	0.5 m	1 m
Gentle slopes or undulations	0.2 m	0.5 m	1.0 m	2.5 m
Very steep	0.5 m	1.0 m	2/2.5 m	5 m

9.3.3   Methods of locating contours

All contouring methods involve heighting a number of points, known as *spot heights*, then deducing the position of the contours from these spot heights. The various methods may be classed as *Direct* or *Indirect*. Direct or 'contour chasing' methods require the location of spot heights actually on the desired contours, at intervals of from 5 to 50 m apart, depending upon the scale of the plan and the shape of the ground. The spot heights are joined by a smooth curve, similar to connecting the points on a graph. Indirect methods entail locating spot heights in positions such that, for any pair of adjacent 'spots', the ground surface may be considered to be a line of constant slope or gradient from one spot height to the next. Contour positions are then interpolated between the spot heights and again the contour points are joined to form smooth curves. Interpolation is generally carried out by estimation rather than with any great precision, e.g. if two spot heights are A and B, with heights of 34.8 and 35.4 m respectively, then the 35 m contour would be 2 units from A and 4 units from B and lying on the line connecting A and B. In this case, the contour position is one-third of the way from A towards B, the third being estimated on plan.

The actual method to be used depends upon a number of factors, including whether the task is to contour an existing plan or involves the survey of both detail and heights. Again, the plan scale and the general form of the ground are important.

9.3.3.1   Direct contouring

These methods are seldom used, except in special circumstances. They should be the most accurate, but are generally slow and hence more costly than indirect methods. Direct methods are at their best on scales of 1:500 to 1:1000 and when the slope and shape of the ground are clearly visible to the surveyor.

(i) Direct contouring by arbitrary selection of spot heights along the contour. The following example illustrates the method.

Imagine a level set up and levelled from a BM, the collimation height of the instrument being found to be 104.230 m and it is required to locate the 103 m contour line. It will be clear that when the base of the staff is standing on the 103 m contour line, then the centre hair reading on the staff will be $104.230 - 103.000 = 1.230$ m. To locate a point on the contour, the staff holder should hold the staff erect, base on the ground and then be directed up or down hill until the centre hair reads 1.230 m. When the reading is correct, a peg or lath may be placed at the staff position to mark a point on the contour. Repetition of this process, the staff holder moving along the slope between each point fixing, will establish a series of points on the contour. Other contour lines may then be fixed in the same way, with the appropriate centre hair reading for the particular contour being selected. To facilitate the work, a line of levels should be run through or around the site initially, giving a series of TBMs such that, wherever the surveyor sets up, there will be a TBM visible and the collimation height may be readily obtained. Note that the spot heights should extend somewhat beyond the site boundaries, to avoid distortion of the contours at the boundaries.

The plan positions of the contour line points must be located thereafter and this can be done by any appropriate detail supply method. If there is no plan, then it is customary to survey both the site and contour points by chain survey. The various contours may be identified by different coloured laths and when the levelling is complete the contour lines may be picked up from the chain lines as other detail.

If a plan already exists, then it may be better to fix the position of the spot heights by bearing and distance from the instrument station. This requires a level fitted with a horizontal circle for the bearings, the distances to the staff being obtained from the stadia hair readings and the instrument stations must be located by measurements from site detail so that they may be plotted on the plan.

(ii) Direct contouring by sections. In this method, a base line is laid out on the site approximately parallel to the contours, then a series of ordinates or section lines set out at right-angles to the base line. Thereafter, contour points are located on each section line, in the same way as described above, effectively fixing the points where each contour line crosses all the section lines. In some circumstances this approach may be easier and quicker than the basic method.

9.3.3.2   Indirect contouring

These methods are the most popular for area levelling, whether by level and staff or by trigonometric heighting. The level and staff should be used for scales of 1:1000 or larger and trigonometric heighting with theodolite stadia readings at smaller scales. Modern sophisticated bearing and distance equipment can be used at all scales, see Part 4.

(i) Grid levelling. This is the most commonly used method when the site is not extensive and has only minor variations in slope. The area is covered by a grid of lines, generally forming squares of 5 to 30 m or so, levels being taken at the intersections of the grid lines. It is common practice to give the lines in one direction identifying letters (A, B, C, . . .) and in the other direction identifying numbers (1, 2, 3, . . .), thus any intersection point may be specified, e.g. A4, B6, C3, etc.

The grid spacing will be influenced by the extent of the undulations of the ground, the accuracy required and the need to be able to draw a

reasonably accurate smooth curve through the interpolated points. The
last point tends to limit the maximum grid spacing on the plan to 25 or
30 mm, that is 5 or 6 m apart on the ground at 1:200 scale, 12 to 15 m at
1:500 and 25 to 30 m at 1:1000 scale. Generally the grid is laid out by tape
and optical square, offsetting particularly on small sites.

The grid must be tied to a straight line, often one of the chain lines is
used, but any straight part of detail may be suitable. If the offsets are to
be kept short, then it is preferable that the grid be based on a straight
line through the centre of the site. Often offsets will seem excessive by
chain survey standards, but errors which arise should not significantly
affect the end product. Positions for spot heights (intersections) may be
ground marked with pegs, short pieces of lath, or arrows with strips of
bunting tied to the loops. On larger sites, ground marks may be
dispensed with by using ranging rods around the perimeter, or pairs of
rods on two adjacent sides, the staff holder being responsible for taping
distances between the rods to locate spot height positions.

Where there are marked changes of slope lines through the site, it is
advisable to attempt to arrange for grid lines to pass along these
features. Alternatively, additional spot heights may be needed and
means of identifying them. Again, the spot heights should extend
beyond the site boundary, but if this cannot be arranged then additional
spot heights at the boundaries will be required. The simple method of
letter/number for spot identification may not always be convenient. An
alternative method is to identify each point as a distance left or right
from a known distance along a central grid line, as in *Figure 9.2*. This
technique is also used for booking sections and cross sections.

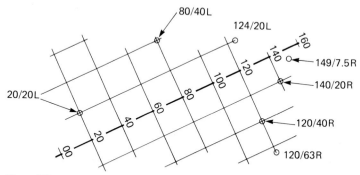

Figure 9.2

(ii) Contouring by sections at changes of slope. This method is similar to
the second of the direct methods, using a base line and ordinates or
sections at right-angles to it. In this case, however, the staff holder goes
along each ordinate to a change of slope point rather than a contour
point, the spot heights being identified as in *Figure 9.2*.

This method is not as popular as simple grid levelling, perhaps because
it is not so convenient for volume calculations. In addition, if it is to be
economical in time then the decision as to where to take the spot heights
must be made by the staff holder and he therefore needs to be
experienced.

(iii) Contouring by spot heights at arbitrary change of slope points. In this method, spot heights are taken over the whole area of the site at all those points at which the surveyor (or the staff holder) considers there are significant changes of slope. Some positions will be self-evident, e.g. changes of slope at the site boundaries, on ridge or valley lines, at the top of conical hills, at the bottom of hollows, etc. It is more difficult on long convex or concave slopes, but it should be remembered that it is unlikely that spot heights will need to be closer than 6 m apart for 1:500 scale mapping, 12 m at 1:1000 scale, and so on.

The spot height positions will have to be located by the methods of *Figure 9.2*, or by bearing and distance from the instrument position. For this reason, the method is not generally popular with a level unless bearing and distance methods are being used. It is, however, the ideal method for theodolite bearing and distance techniques in conjunction with trigonometric heighting.

9.3.4   Booking and the reduction of levels

The levelling procedure, including the booking, should be similar to that previously described. The 'Remarks' column of the book may need to be modified so that the spot heights can be identified and located. In grid levelling a column could be headed 'Grid point', while in bearing and distance methods columns may be headed 'Hor. Angle', 'Stadias', 'Dist'. When using stadia distances from instrument to staff the distance can be mentally deduced in the field and entered in the Dist column, without writing down the actual stadia readings, but it is better practice to record the stadia hair readings in the book. If sections or ordinates at right angles to a base line are used, then it will be necessary to include extra columns for these.

Reduction should preferably be by the rise and fall method, since there are likely to be many intermediate sights. At an intended 0.5 m contour VI, then readings to the nearest 0.005 m and reduced to 0.05 m are more than adequate and proportionally at other contour intervals.

9.3.5   Interpolation and the plotting of contours

Whichever method has been used for spot heighting in the field, in the office the spot heights are plotted in their correct positions on the plan and the reduced level of the point written alongside, to the limit quoted above. In indirect methods, every adjacent pair of spot heights is examined and, if one is above the level of a required contour line while the other is below, then the actual position of the cut of the contour line is estimated between the spot heights and marked.

Figure 9.3 illustrates a method of interpolation, but it also shows a problem which may arise with a beginner. Here a study of the interpolated points shows that the contours will cross the grid square either at an angle of roughly 45/225° or 135/315°. The square cannot be part of a plane surface, but must contain either a valley running SW–NE or a ridge running NW–SE, and there is no information to indicate which. In practice, a smaller grid could have been used, or a note should have been entered in the field book to the effect that one of the diagonals of the square was a ridge or a valley line.

The interpolated points, or the contour points located by the direct method, should be joined by smooth freehand curves to represent the contour lines. It should be appreciated, however, that the contour line should be as straight as possible between each adjacent pair of common points. Adjacent contour lines are often roughly parallel, and they can only end at a plan edge. The points made in Section 9.3 should also be remembered.

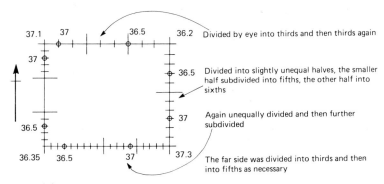

Figure 9.3

If the plotter is satisfied with the pencilled contour lines then they may be inked in. Traditionally contours are coloured brown or orange, but for some reprographic processes they must be in black. To distinguish them from detail and provide contrast, they should be drawn in a narrower gauge line, or in a different form of line. Every fourth or fifth contour line may be emphasized by a wider gauge of line. The height of contour lines is indicated at a gap in the lines, such that they are read looking uphill, and if at all possible they should be capable of being read without having to turn the plan (see *Figure 9.1*).

9.3.6   Other methods of representing relief

There are many methods of representing relief other than by the use of contour lines, the use of slope symbols as in *Figure 2.10* being one. Not many of these other methods are suitable for site plans of proposed construction sites, but some may be useful in special circumstances, e.g. in illustrating the remains of a mediaeval village, where traditional contours and spot heights would be quite inadequate. For further information, see the Bibliography.

9.4   Sections and cross-sections

In building and engineering work it is often necessary to prepare a profile of the ground along a particular line. Such a profile, termed a *section of levels*, is obtained by running levels along the required line, then plotting the heights and distances on paper to appropriate scales. The transfer of a line (straight or curved) from a plan or other setting out document is covered in Part 5, Setting Out.

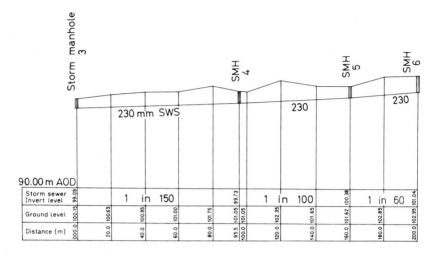

Longitudinal Section: Storm sewer 'A', CHGE. 0 to 2 + 00

Scales: Horizontal 1 : 500
        Vertical 1 : 100 (original plotting scales)

    (All measurements in metres)

Figure 9.4

9.4.1 Longitudinal sections

A ground profile along the centre or other longitudinal line of an existing or proposed road, railway, pipeline, canal, etc., is termed a *longitudinal section*. Levels are observed along the ground line by series levelling (Section 8.6), generally at a standard interval of horizontal distance such as 20, 25, 30, 50 or 100 m, together with points where the ground slope changes distinctly or natural or artificial features disturb the ground profile. As always, to minimize error, levelling should start and finish on BMs, or be closed by flying levels and back and foresight distances be kept equal. It is useful to use changepoints on well-defined features which can be used again on the check-levelling to localize error.

The reduced levels are plotted using distorted scales so as to emphasize height variations. Typical examples are a horizontal scale of 1:500 with a vertical scale of 1:100, or a horizontal 1:2000 with a vertical 1:200. *Figure 9.4* shows a typical example of a longitudinal section of the ground along a proposed sewer line. The same section also shows details of the invert levels of the sewer manholes proposed and the suggested gradients of the sewer pipe between manholes. Note that a vertical line is drawn at every ground point to the scale height and that the ends of these lines are connected by straight lines, no attempt being made to draw the surface as smooth curves, since the scale distortion makes this pointless.

All details as to levels and distances are shown in the 'boxes' drawn across the bottom of the section. Sufficient space must be left between the top line of the boxes and the lowest level of the section to allow for extra boxes which may be required for other details such as, for a road, perhaps road formation levels, horizontal curve data, storm sewer details, foul sewer details, cut and fill for excavation, etc.

Note in *Figure 9.4* that the datum height should preferably be a multiple of five metres and that some 50 to 100 mm clearance is reasonable between the top line of the boxes and the lowest level of the section. When booking levels for this type of work it is sensible to have a 'Distance' column within the remarks space in the book.

9.4.2   Cross sections

Where the proposed construction is of considerable width the longitudinal section information must be supplemented by cross-sections. A *cross-section* is a profile of the ground at right-angles to the longitudinal line, serving mainly to allow the calculation of the volumes of earthworks. Cross-sections are not usually taken for pipelines, but are required for roads, railways and canals. Cross-sections must be taken at regular intervals along the centre-line, the same regular points generally as used for the longitudinal section. The distance apart depends upon the nature of the ground, perhaps 20 m on broken ground, or even 100 m on gentle slopes. Occasional sharp changes in ground configuration, such as rock outcrops, may necessitate extra sections at other non-regular points. The centre-line should be pegged at all points where cross-sections are to be taken, the pegs being driven to ground level and marker pegs placed beside them for identification. The pegs are best placed before the longitudinal section is taken, then the peg levels provide a comparison between long-section and cross-section levels if the two tasks are done separately, cross-section levelling acting as the check-levelling for the long-section levelling. Alternatively, both long-section and cross-sections may be levelled at the same time and checked by flying levels.

A cross-section is identified by the longitudinal section chainage at its centre and distances on the cross-section are measured and noted as left or right of centre-line.

This is one of the methods recommended for use in Indirect Contouring. The width of a cross-section is fixed by consideration of the construction width and the width of land reserve available. The distances left and right are measured by synthetic tape, directions usually being judged by eye, like offsets in chaining. Levels are taken at the centre-line, at all changes of slope and at the extreme width of the section. Flat ground may only need three levels, broken ground perhaps twenty or more.

The levelling is normally series-levelling and usually one cross-section is completed at a time, then on to the next, and so on. Very steep side-slopes may need two or more set-ups per section and then it may be faster to take the downhill levels for two or more sections from one set-up and their uphill levels from another set-up. Booking must be done very carefully in this case, to avoid mixing the levels of the two sections. The levelling and chainage measurements may also be successfully carried out by bearing and distance techniques as in Chapter 14.

When plotting cross-sections, it is normal to use the same scale both horizontally and vertically. Generally the scale is that used for the verticals on the longitudinal section for the same job. In the two examples quoted earlier, the cross-sections would probably be plotted at 1:100 and 1:200 respectively (see *Figure 9.5*), but note that 29 + 20.0, 29 + 40.0, etc., is the chainage notation system used for roads and railways.

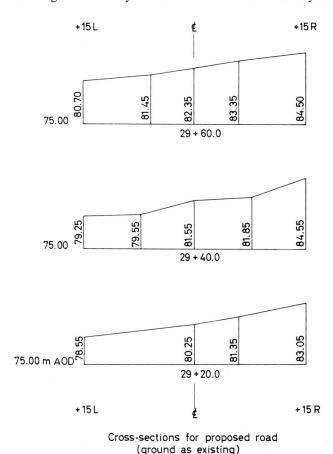

Cross-sections for proposed road
(ground as existing)

Original scales: Horizontal and vertical 1: 200
(All measurements in metres)

Figure 9.5

9.5 Precise levelling

As stated previously, geodetic levelling such as the heighting and adjustment of large levelling nets is not within the scope of this text, but it may be necessary on occasion to transfer levels with very high accuracy.

For such work a *precise tilting* or *precise automatic level* with *parallel plate micrometer* should be used, in conjunction with *invar levelling staves* (see Sections 8.5.2 to 8.5.4).

Invar staves may be single-scale or double-scale, the former appearing to be the most popular in the UK. Double-scale staves generally have the two sets of graduations on the same face, but some models have them on the opposite faces. The two scales have their graduations offset by a fixed amount.

9.5.1 General procedure for precise levelling

The instrument should be allowed to settle at the local air temperature before commencing work. Select firm changepoint positions, arranged so that back and foresight distances are equal to within 0.5 m or so, either by taping or by pacing the distances out carefully. If the difference of the sum of the lengths of the foresights and the sum of the lengths of the backsights exceeds 1.5 m during the levelling, then future instrument positions must be arranged to reduce the difference. Observing distances should be from about 25 to 40 m and grazing rays within about half a metre of the ground should be avoided due to the effects of variable refraction of the air. The level should be protected against sun and wind by umbrella and windshield, if necessary, but levelling should not be carried out in high winds.

Using two staves, one should be placed at all odd staff points and the other used for all even staff points, to ensure that all observations at one point are made on the same staff. At odd instrument positions, commence observations with a backsight; at even instrument positions commence with a foresight, to help reduce systematic instrument errors. When setting up the instrument at each station, point the telescope at the staff on which the first reading is to be taken, before levelling-up the circular bubble.

For best results, the line should be levelled at least twice; once outwards, once backwards in the opposite direction, and the two sets at different times of day under different atmospheric conditions, using different changepoints. The results of the two levellings will be meaned if their difference is within acceptable limits. Note that although the instrument and staff positions are set out by measurement, the actual distance to the staff from the instrument must be noted at each position using the stadia hairs. The line should be arranged so that there are an even number of instrument set-ups. The actual observing order depends upon whether single-scale or double-scale staves are used.

9.5.2 Using single-scale staves

Referring to *Figure 9.6*, successive instrument stations are numbered 1, 2, 3, etc. It is required to level from BM A to a distant point E. B1 means 'Backsight reading from instrument station 1', while F1 means 'Foresight reading from instrument station 1'.

Staff readings are made in the sequence:

B1, F1; (from stn 1)
F2, B2; (from stn 2)
B3, F3; (from stn 3), etc.

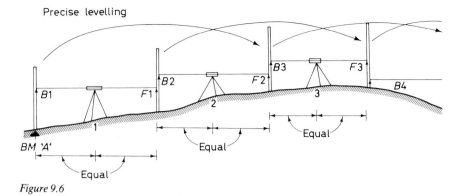

Figure 9.6

At each observation, read and book all three hair readings; the central and the two stadia hairs, using the parallel plate micrometer for each in turn, see Section 8.6.4. For example

	Readings to B1	Reading order		Readings to F1
Centre hair reads	1.895 49 m	2	3	1.539 52 m
Upper stadia hair	2.057 10	1	4	1.702 95
Lower stadia hair	1.733 62	1	4	1.376 45
Centre hair reads	1.895 37	6	5	1.539 43
Sum stadia rdgs	3.790 72			3.079 40
Sum centre rdgs	3.790 86			3.078 95
Diff of stadia rdgs (×100)	32.348 m			32.650 m

Note: Throughout this book, *lower hair* or *bottom hair* is taken to mean the stadia hair or line which indicates the smallest reading on the observed staff.

Before proceeding further, the following recommended limitations should be checked:

Centre hair readings on the same staff should agree within 0.00050 m.
The sums of the readings on to the same staff to agree within 0.00100 m.
The horizontal distances B1 and F1 to agree within 0.5 m or so.

If these limitations are not achieved, the observations should be repeated as necessary, the objective being to reduce, as much as possible, the effect of cumulative errors.

To compute the total difference in level between A and E, total the backsights, total the foresights and their difference gives twice the required amount. To check the arithmetic, compute also the rise or fall between every pair or changepoints, the difference of the totals must equal the previous figure.

This may be expressed as:

$$H_E - H_A = \Sigma(B - F) = \Sigma_B - \Sigma_F.$$

9.5.3   Using double-scale staves

The procedure is similar to that used with single-scale staves, but instead of the centre hair being read twice, each centre hair is read once on each of the two scales. Both sets of stadia hairs are usually read and booked. The reading checked are obtained by comparing the readings on the two scales, centre readings against centre readings, stadia differences or distances against stadia differences or distances. The same limitations are recommended and if these are not achieved the observations must be repeated, otherwise proceed.

Finally, the difference of elevation of E and A is obtained as before, but two values are obtained from the two sets of scale readings. The two values should agree within a suitable tolerance, for example $3\sqrt{K}$ mm in which K is the distance levelled in kilometres. If the line is levelled twice as is normal practice, then the two final results are meaned again.

9.6   Reciprocal levelling

Where it is not possible to equalize the backsight and foresight distances from an instrument position, such as in taking levels across a wide river, the technique of *reciprocal levelling* may be used.

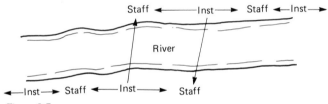

Figure 9.7

9.6.1   Distances of 60 to 100 m

In this instance, the method is as for ordinary levelling, except that the gap should be observed twice, as in *Figure 9.7*.

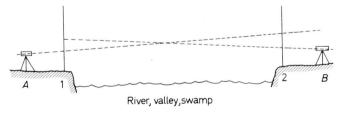

Figure 9.8

9.6.2   Longer distances, ordinary work

In *Figure 9.8*, the level is first placed at position A, then at position B, and level staves at positions 1 and 2. The distances from A to 1 and from B to 2 should be equal. From A, readings are taken to 1 and 2, and from B readings are taken to 1 and 2. The readings are noted as A1, A2, B1, B2 and the differences (A2 − A1) and (B2 − B1) calculated. These differences will not be the same, but their mean gives the difference in level between points 1 and 2, provided the atmospheric conditions have not changed between taking the two sets of readings.

Should there be a long time delay in getting the instrument from one side of the river to the other, then it is better to use two levels simultaneously, to overcome the possibility of changing refraction. A target on the staff may be necessary as distances become very large.

9.6.3   Longer distances, precise work

In precise work, the traditional approach was to use a precise level with either a gradienter screw or a bubble scale for distances between 100 and 1500 m. Beyond 1500 m the preferred method was *reciprocal trigonometric heighting* with two theodolites which could be read to decimals of a second, many FL and FR readings being needed. Automatic levels and most modern tilting levels, are not fitted with gradienter screws or bubble scales, thus, for all precise work over 100 m the reciprocal trigonometric heighting technique is recommended.

Part 3

Areas and volumes

Chapter 10

Area calculations

10.1 Introduction

The surveyor is often required to determine areas of land and sometimes to sub-divide areas and fix new boundaries. Volume calculations, e.g. in construction earthworks, disused quarries, opencast mining, etc.; all involve a knowledge of area calculation methods first.

10.1.1 Area formulae for simple figures

10.1.1.1 Rectangle, square

Area = length × breadth$\qquad\qquad$(10.1)

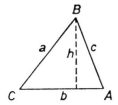

Figure 10.1

10.1.1.2 Triangle

Given base length b and perpendicular height h,

Area = $b \times h/2$$\qquad\qquad$(10.2)

Given sides a, b and c and $s = (a + b + c)/2$,

Area = $\sqrt{s(s - a)(s - b)(s - c)}$$\qquad\qquad$(10.3)

Given two sides, a and b and included angle C,

Area = $(a\,b\,\sin C)/2$$\qquad\qquad$(10.4)

Given side a and angles A, B and C,

Area = $(a^2 \sin B \sin C)/(2 \sin A)$$\qquad\qquad$(10.5)

219

10.1.1.3   Trapezium

Given parallel sides a and b and perpendicular distance between them h,

$$\text{Area} = (a + b)\, h/2 \tag{10.6}$$

10.1.1.4   Circle

Given radius r and $\pi = 3.141592$,

$$\text{Area} = \pi r^2 \tag{10.7}$$

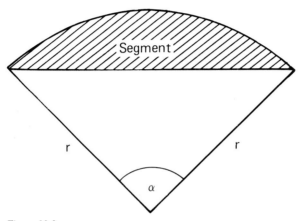

Figure 10.2

10.1.1.5   Segment of a circle

Area of the sector less the area of the triangle in *Figure 10.2*,

$$\text{Area} = (\pi r^2\, \alpha\, /360) - (r^2 \sin\alpha)/2 \tag{10.8}$$

Note that $(r^2 \sin\alpha)/2$ comes from Equation (10.4).

10.1.2   Area units

The standard area units are the square kilometre (km), the hectare (ha) and the square metre (m^2). All may be used in surveying as appropriate.

$$10\,000\,\text{m}^2 = 1\,\text{ha (approx. 2.5 acres)}$$
$$100\,\text{ha} \quad = 1\,\text{km}^2$$
$$= \text{the area of one O.S. National Grid 1:2500 map.}$$

10.1.3   Area and boundary terms

For calculation purposes, the boundaries of areas on plans and drawings may be made up of straight lines, a *rectilinear boundary*, or of irregular lines, known then as *curvilinear boundaries*. Where the boundary of an area is curvilinear, as in *Figure 10.3*, straight *give and take lines* may be drawn on the plan to replace the actual boundaries for calculation purposes. These lines should be placed so that the areas excluded by them are approximately equal (by eye) to the external areas taken in by them. When the boundaries have been 'averaged out' in this way the figure becomes rectilinear and the area can be calculated from triangles and trapeziums.

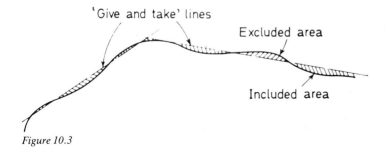

Figure 10.3

Land owners occasionally have a curvilinear boundary replaced on the ground by a rectilinear one, to allow more effective use of the land. If the area of the parcel either side of the boundary remains the same, then the boundary is known as a *give and take line*.

10.2   Areas from plotted plan or other drawing

Areas may be obtained from drawings, from field measurements (with or without plotting) or from calculated data such as rectangular co-ordinates or partial co-ordinates. In some cases the methods may overlap, but the logical procedure is to determine the data available and then select an appropriate calculation method.

10.2.1   Sub-division into triangles

Any straight-sided or rectilinear figure may be sub-divided into triangles by drawing appropriate straight lines. The figure area is then the sum of the areas of the triangles. The method may be used on a drawing, or the necessary measurements may be made in the field and the area calculated directly. On occasion it may be possible to use trapeziums, this slightly reducing the labour involved.

10.2.2   Counting squares or dots

The *squares method* uses a transparent or translucent overlay, with a grid drawn on the overlay, each grid square representing a unit of square measure. A typical type is 2 mm squared paper with each tenth line in a heavier gauge. At 1:500 scale each small square represents $1\,m^2$ and each large square $100\,m^2$. At 1:2500 scale the areas would be $25\,m^2$ and $0.25\,ha$ respectively.

The transparency is placed over the area to be measured so that as many large squares as possible fall within the boundary and a line of the grid made to coincide with one or more of the rectilinear boundaries, if any. The squares are now counted, part squares at the boundary being counted as either in or outside the parcel as judged by eye, see *Figure 10.4*, the hatched area being counted.

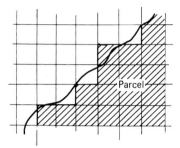

Figure 10.4

The grid should then be shifted and rotated in plan and a fresh count made. Preferably this should be done still again and the results compared. An area correct to 0.5% should be expected.

An alternative method, considered by some to be a better method, is to replace the grid of squares by a grid of dots at the same spacing as the grid lines, then count the dots included in the figure. Either method is suitable if there are many areas to measure, but it is uneconomical for a single parcel of land if the grid has to be specially produced by hand drawing.

10.2.3   Ordinates methods

10.2.3.1   Ordinates overlay

This method uses an overlay like the square counting method, but the overlay is ruled with equally spaced parallel ordinates, their distance apart being some convenient interval of, say, 8 or 10 mm. The ruled sheet is placed over the area to be measured and estimated give and take lines drawn in as in *Figure 10.5*. Knowing the distance between the ordinates, to scale, the distance between the vertical give and take lines may be scaled off for each block and the area of each block calculated as a simple rectangle. The sum of the areas of the rectangles will give an estimate of the area of the whole figure.

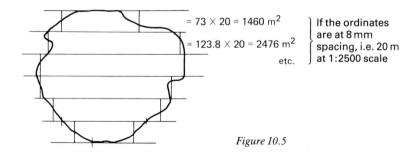

= 73 × 20 = 1460 m²

= 123.8 × 20 = 2476 m²

etc.

If the ordinates are at 8 mm spacing, i.e. 20 m at 1:2500 scale

Figure 10.5

10.2.3.2   Mean ordinate rule

In this method, a line is drawn through the centre of the area to be measured, as in *Figure 10.6*, the line length being d. The line is divided into equal intervals, of length l and ordinates drawn at right-angles to the line, the length of the ordinates being scaled as o1, o2, o3, . . . on.

The *mean ordinate rule* states that the area is equal to the mean ordinate length by the total length of the line.

$$\text{Area} = (d/n)(\text{o1} + \text{o2} + \text{o3} + \ldots + \text{o}n) \qquad (10.9)$$

where d is the total line length, n is the number of ordinates and the ordinates are o1, o2, o3, . . . The method is rapid, but not very accurate.

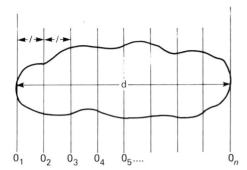

Figure 10.6

10.2.3.3   Trapezoidal rule

In this method, the shape formed between each pair of ordinates is considered to be a trapezium, then summing the area of each trapezium gives

$$\text{Area} = (l/2)(\text{o1} + 2\text{o2} + 2\text{o3} + 2\text{o4} + \ldots + 2\text{o}(n-1) + \text{o}n) \qquad (10.10)$$

If the boundary is curvilinear then the area is a good approximation and the accuracy may be increased by increasing the number of ordinates.

10.2.3.4   Simpson's rule

This is very similar to the trapezoidal method, but assumes that the irregular boundary consists of a series of parabolic arcs between the ordinates rather than straight lines. In this case, however, the area must be divided into an *even* number of strips by an *odd* number of ordinates. If there are an odd number of strips, then the last strip area must be calculated separately and added to the area calculated for the even number of strips.

$$\text{Area} = (l/3)(\text{o1} + 4\text{o2} + 2\text{o3} + 4\text{o4} + \ldots + 2\text{o}(n-2) + 4\text{o}(n-1) + \text{o}n)$$
$$(10.11)$$

The rule may be stated in words as:

Where an area with curvilinear boundaries is divided into an even number of strips of equal width, the total area is equal to one-third the strip width multiplied by the sum of the first and last ordinates, twice the sum of the remaining odd ordinates and four times the sum of the even ordinates.

10.2.4   The planimeter

The *planimeter* is a mechanical device for integration, i.e. the calculation of the area under the curve of a function. Area measurement from drawings by planimeter is the most efficient and fast method, particularly if the areas are small or very irregular in shape. The forms used in survey are known as *polar planimeters*, their accuracy typically being similar to that for counting squares, 0.5%. Recent electronic models are probably more accurate.
   The elements of the instrument are illustrated in *Figure 10.7*.

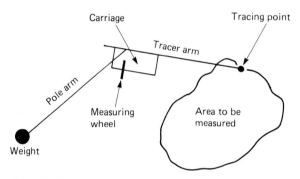

Figure 10.7

(1) The *pole arm*, with a needle pointed weight at one end, the *pole weight* sometimes being separate. The other end of the pole arm carries a pivot resting in a socket in the tracer arm.
(2) *The tracer arm*, fitted at one end with a *tracing point* with an adjustable support.
(3) The *carriage*, which may be fixed to, or may slide along, the tracer arm, has a *measuring wheel* and a *counting dial*.
(4) The *counting dial* records the number of revolutions made by the wheel, which may also be fitted with a vernier scale.

   Planimeters with fixed carriages are known as *fixed arm polar planimeters*, and will record the area either in square inches or square centimetres only. *Sliding bar polar planimeters* (the carriage may slide along the tracer arm) have a scale on the tracer arm so that the carriage may be set at some particular map scale so that one revolution of the wheel will equal one specific unit of area. For example, at 1:1250 scale, one revolution of the measuring wheel is equal to one hectare, or at 1:100 scale, $100 \, \text{m}^2$.

The simplest application of the planimeter is as follows. Set the carriage as necessary for the scale of the drawing, attach the pole, pole arm to the tracer arm. Place the needle pointed weight outside the area to be measured, with the pole and tracer arms roughly at right-angles to one another and the tracing point approximately in the centre of the area. Check, without moving the weight, that the tracing arm can reach any part of the figure boundary without the measuring wheel moving across the edge of the drawing material. If this cannot be done, try the pole weight in a different position or divide the area into several smaller areas. Locate and mark a starting point on the boundary, an ideal position being a 'dead' point where a slight movement of the tracing point causes no rotation of the wheel. Set the tracing point on the starting point, on older models note the reading, on newer models set the scale to zero. Carefully move the tracing point clockwise around the boundary, terminating exactly on the starting point, note the number of revolutions on the dial and the part revolutions on the measuring wheel drum and vernier as appropriate. Subtract the start reading, if any, then convert the result to an area in hectares or square metres.

In practice, the operation is rapid, but results depend upon the operator. The procedure should be repeated several times, obtaining several values for the area and when three consistent results have been achieved they may be meaned and the mean value accepted. An area can be measured with the pole inside the figure, but the result is less accurate and this should be avoided by dividing the area into smaller areas.

Although planimeters are reasonably robust they should be handled with care, particularly the measuring wheel – the edge should not be touched by hand or brought into contact with hard surfaces. A checking rule or test bar should be carried in the instrument case, to test the instrument's accuracy. A check may also be made by circumscribing an area of known size. Planimeters may develop a fault known as *non-parallelism*, i.e. the axis of the measuring wheel not parallel to the tracer arm. This may be checked by measuring an area with the weight to the left, then again with the weight to the right of the area to be measured. A correction factor may be deduced for the correction of subsequent area measurements.

A recent model is the Tamaya Planix 7, with keyboard and LCD display. At the keyboard may be entered the survey scale, or the horizontal and vertical scales if working from sections, the required unit for results (m^2, km^2, acres, etc.) and whether the user wishes to average a number of results. An accuracy of $\pm 0.2\%$ or better is claimed.

10.3   Areas from survey field notes with no plan

On occasion the area of a piece of land is required, but no plan, e.g. land or crop compensation, crop areas for subsidy, etc. Survey lines will normally divide the survey area into triangles or trapeziums and then the formulae (10.2) to (10.6) may be used to calculate the survey area, direct from field bookings, without plotting a plan. If the field bookings are tied to a closed loop traverse it may be preferable to obtain the co-ordinates of the traverse then calculate the area as shown below.

Where there are narrow strips of land outside the survey triangles, offsets may be taken to act as ordinates and Simpson's or the Trapezoidal rule used to calculate the strip area. This is not generally practicable in a survey which is to be plotted, since in that case such offsets are usually taken at changes of direction of the detail or boundary.

10.4   Areas from co-ordinates

If a closed traverse has been co-ordinated, or the boundary of a property or a cross-section is defined by co-ordinated points, then it is possible to calculate the area of the figure concerned directly from the co-ordinates without plotting the figure on paper. *Figure 10.8* shows a four-sided closed loop traverse. It will be evident that the area of the figure ABCD is equivalent to the addition and subtraction of trapeziums:

$$\text{Area} = (N_B - N_A)(E_A + E_B)/2 + (N_C - N_B)(E_B + E_C)/2$$
$$- ((N_C - N_D)(E_C + E_D)/2 + (N_D - N_A)(E_D + E_A)/2)) \qquad (10.12)$$

or

$$2 \times \text{Area} = E_A \cdot N_B - E_A \cdot N_A + E_B \cdot N_B - E_B \cdot N_A + E_B \cdot N_C - E_B \cdot N_B + E_C \cdot N_C - E_C \cdot N_B - E_C \cdot N_C + E_C \cdot N_D - E_D \cdot N_C + E_D \cdot N_D - E_D \cdot N_D + E_D \cdot N_A - E_A \cdot N_D + E_A \cdot N_A$$

or

$$2 \times \text{Area} = (E_A \cdot N_B + E_B \cdot N_C + E_C \cdot N_D + E_D \cdot N_A) - (N_A \cdot E_B + N_B \cdot E_C + N_C \cdot E_D + N_D \cdot E_A) \qquad (10.13)$$

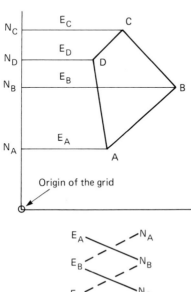

Figure 10.8

10.4.1   The Quick method

Equation (10.13) is sometimes known as the *quick method*, and in use the co-ordinates are simply listed one above the other, with the co-ordinates of the first point being repeated at the end of the list, as shown in *Figure 10.8*. Twice the area is equal to the sum of the products of the solid lines less the sum of the products of the pecked lines.

10.4.2   The cross co-ordinate method

Where the co-ordinates must be keyed by hand into a calculator, this method may be preferred since it can provide a check.
    Equation (10.12) may be re-written simply as

$$2 \times \text{Area} = \Sigma E_{AB}. \triangle N_{AB} + \Sigma E_{BC}. \triangle N_{BC} + \Sigma E_{CD}. \triangle N_{CD} + \Sigma E_{DA}. \triangle N_{DA} \tag{10.14}$$

Similarly it can be proved that

$$2 \times \text{Area} = \triangle E_{AB}. \Sigma N_{AB} + \triangle E_{BC}. \Sigma N_{BC} + \triangle E_{CD}. \Sigma N_{CD} + \triangle E_{DA}. \Sigma N_{DA} \tag{10.15}$$

Equations (10.14) and (10.15) provide the solution and check, one answer being positive and the other negative. The data may be listed as in *Table 10.1*.

TABLE 10.1.

Point	'E'	'N'	
A	100.00	100.00	
B	133.99	321.43	
$\triangle$	33.99	221.43	Sum $\Sigma E^s \cdot \triangle N^s$
Σ	233.99	421.43	$= -78\,699.6474\,\text{m}^2$
B	133.99	321.43	
C	326.09	294.80	Sum $\triangle E^s \cdot \Sigma N^s$
$\triangle$	192.10	-26.63	$= +78\,699.6474\,\text{m}^2$
Σ	460.08	616.23	$= 2 \times \text{Area}$
C	326.09	294.80	
D	280.96	99.97	$\therefore \text{Area} = 39\,349.8237\,\text{m}^2$
$\triangle$	-45.13	-194.83	$= 3.935\,\text{ha}$
Σ	607.05	394.77	
D	280.96	99.97	
A	100.00	100.00	
$\triangle$	-180.96	0.03	
Σ	380.96	-199.97	

10.4.3   The double longitude method

This method is slower than the others, but has the advantage that the area can be calculated directly from the partial co-ordinates of the lines of the traverse. The rule is that the area is the algebraic sum of the products of the latitude of each line and the longitude of that line. The terms *latitude* and *longitude* are obsolete in the context of rectangular co-ordinates, but in this application the *latitude* of a line is defined to be its *partial northing*, while its *longitude* is the *mean eastings value of the line* with respect to some reference ordinate which may, or may not, pass through the origin of the grid. The rule is another re-statement of Equation (10.12). The double longitude of each line can be calculated direct from the differences in eastings.

If the reference ordinate for longitude passed through station A instead of the origin of the grid as in *Figure 10.8*, then the double longitude of the line AB would be $E_B - E_A$. It can be shown that the successive double longitudes are equal to the previous double longitude + the previous $\triangle E$ + the present $\triangle E$. *Table 10.2* shows a typical calculation, using the data from the previous table.

TABLE 10.2.

Line	$\triangle N$	$\triangle E$	Double long	$\times \triangle N$
AB	221.43	33.99	33.99	
BC	− 26.63	192.10	260.08	
CD	−194.83	− 45.13	407.05	
DA	0.03	−180.96	180.96	$= -78\,699.6474\,\text{m}^2$
				$= 2 \times \text{Area}$
			$\therefore\ \text{Area}$	$= 39\,349.8237\,\text{m}^2$
				$= 3.935\,\text{ha}$

The calculation of the double long, and of twice the area is as follows:

$\triangle N$	$\triangle E$	Previous	Previous double long	
$\triangle N_{AB}$	$[(E_B - E_A)]$			$= \triangle N \times \text{Double long}$
$\triangle N_{BC}$	$[(E_C - E_B) +$	$(E_B - E_A) +$	$(E_B - E_A)]$	$= \triangle N \times \text{Double long}$
$\triangle N_{CD}$	$[(E_D - E_C) +$	$(E_C - E_B) +$	$(E_C + E_B - 2E_A)]$	$= \triangle N \times \text{Double long}$
$\triangle N_{DA}$	$[(E_A - E_D) +$	$(E_D - E_C) +$	$(E_D + E_C - 2E_A)]$	$= \triangle N \times \text{Double long}$

$$\text{Or twice the area} = \triangle N_{AB}(E_B + E_A - 2E_A)$$
$$+ \triangle N_{BC}(E_C + E_B - 2E_A)$$
$$+ \triangle N_{CD}(E_D + E_C - 2E_A)$$
$$+ \triangle N_{DA}(E_A + E_D - 2E_A) = 2A$$

$\triangle N \times$ double longitude may be calculated and added progressively on a hand calculator to obtain twice the area.

The $\triangle$Ns $\times (- 2E_A)$s cancel out to leave Equation (10.14), or $2 \times \text{Area}$ $= \triangle N_{AB} \cdot \Sigma E_{AB} + \triangle N_{BC} + \Sigma E_{BC} + \triangle N_{CD} \cdot \Sigma E_{CD} + \triangle N_{DA} \cdot \Sigma E_{DA}$. Note a check of the double longitude calculations, the final double longitude is equal to the final $\triangle E \times (- 1)$.

10.5   Alteration and subdivision of areas

Typical area management problems concern the replacement of one boundary by another (e.g. substituting a straight boundary for a curvilinear one) and the subdivision of areas of land. Existing boundaries may be described loosely as either being in the English system of general boundaries, a boundary being *mered* to a line on a map, or else being defined mathematically by a series of co-ordinated points, as in the fixed boundaries of the Australian Torrens system.

10.5.1   Methods for replacing a curvilinear boundary by a give and take line

10.5.1.1   *Geometrical (graphic) methods*

Figure 10.9 illustrates a boundary plotted on a drawing, the boundary consisting of three straight lines. It is required to replace these by a single straight line. The areas must, of course, be plotted to a convenient large scale.

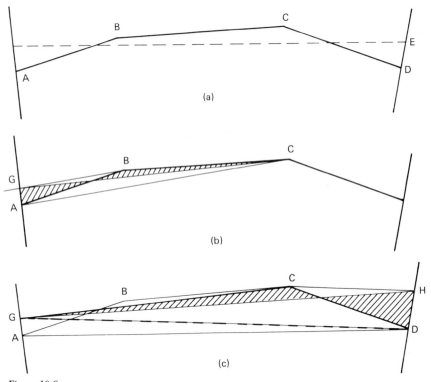

Figure 10.9

Method 1: Lightly draw an estimated give and take line, as in *Figure 10.9(a)*, then by counting squares, calculate the areas ABCDA and ADEFA. If the two areas are equal, the give and take line is in the correct postion. If the ares are not equal, move the estimated line and repeat the area calculations, repeating the process as needed.

Method 2: Draw a trial and error give and take line, as in (1), but measure the area of the figure AFEDCBA, clockwise from A, passing through the points in that order, using a planimeter. If the give and take line is in the correct position then the total area of the figure will be zero, the included areas equalling the excluded area. If not equal to zero, move the give and take line and try again.

Method 3: In *Figure 10.9(b)*, construct the line BG parallel to CA, then the area of triangle ABC is equal to the area of triangle AGC. (The geometrical theorem that the areas of triangles on the same base and between the same parallels are equal.) Construct the line CH parallel to the line GD, then similarly the area of triangle GCD is equal to the area of triangle GHD. The area ABCDA in *Figure 10.9(c)* is now equal to the area AGHDA, thus the line GH is the required give and take line.

10.5.1.2 Numerical methods

Alternatives to graphical solution include direct field measurement and setting out or calculation from co-ordinates.

Method 1: Field method. *Figure 10.10* shows a field sketch of an existing fence to be replaced by a give and take line. Select a point A on the fence from which the new line will commence, then run a chain line from A to any point B on the opposite boundary. From chainage and offset measurements, calculate the area enclosed between the existing fences and the chain line AB.

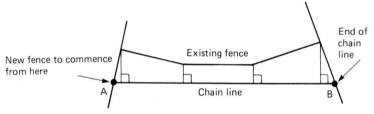

Figure 10.10

Calculate the height (*h*) of a triangle, of area equal to that enclosed area and with a base equal in length to AB. Set out a line parallel to the chain line AB and at a distance of *h* from the chain line. The point where this new parallel line cuts the right-hand boundary defines the other end of the required give and take line and this should be joined to point A.

Method 2: Co-ordinates. *Figure 10.11* illustrates existing boundaries defined by the co-ordinated points A, B, C, D, W, X, Y and Z. It is required to replace the boundary ABCD by a single give and take line.

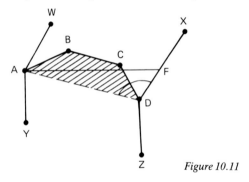

Figure 10.11

Draw the line AD, and calculate the hatched area ABCD. Calculate the length DF, F being a point on the line DX, such that the area of the triangle ADF is equal to the hatched area. The length AD and the angle ADF may be calculated from the known co-ordinates, then Equation (10.4) may be used to determine DF. The line AF is the required give and take line.

10.5.2   Subdivision of areas

Subdivision problems may involve dividing a plot of land into proportional parts, or parts of a specific size or shape. New boundaries may be required to be parallel to some feature, or to start and/or end on given points. Each task may be different, with different methods of solution and setting out and for some, graphical methods will suit, while others will need numerical methods. Setting out problems are covered in Chapter 17.

While no set methods can be laid down, the following general approach is recommended:

For each subdivision plot, ascertain relevant details of shape, location, size, fixed points or limits.

Obtain all relevant data, co-ordinates, plans if any.

Visit the site, carry out recce.

If the size to be set out is a proportion of the whole, survey the whole site, calculate the area, determine the area to be set out.

If the size is specified, decide from the given information which are the known boundary positions and, roughly, where the new boundary will lie.

Insert pegs or other marks on your estimate of where the new rectilinear boundary or boundaries shall be located.

Survey the site as bounded by the existing features and the new pegs. Obtain the area of the new subdivision, this is unlikely to be correct but will provide a start basis.

Deduce how much the new peg or pegs must be moved to obtain the correct area.

Move the pegs, check the area, repeat as necessary.

If the area is to be a regular shape, it may be possible to set it out directly without an intial pegging of the site.

If the area is defined by rectangular co-ordinates, the new boundary points may be computed before visiting the site, using the formulae given earlier.

10.6   Microcomputer applications

The MICROSURVEY Trigonometry program includes routines for all the standard shapes and surfaces likely to be encountered, together with cross-section cut and fill area and also the Trapezoidal and Simpson's rules.

It must be remembered that this program does not provide data storage or amendment facilities, but a hard copy of the screen display for the individual problem may be obtained.

Chapter 11

Volume calculations

11.1 Introduction

Volume calculations in surveying are generally concerned with the volume of earthworks and may be subdivided into calculations based on cross-sections, used for roads, trenches, etc., and calculations based on contours or spot heights.

11.1.1 Volume formulae for simple figures

11.1.1.1 Cube, prism, cylinder

Given base area A and height h,

Volume $= A \times h$ (11.1)

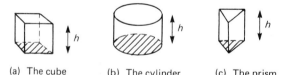

(a) The cube   (b) The cylinder   (c) The prism

Figure 11.1

11.1.1.2 Wedge

Given base area A and height h,

Volume $= A \times h/2$ (11.2)

Figure 11.2

232

11.1.1.3   Pyramid, cone

Given base area A and height h,

  Volume $= A \times h/3$ (11.3)

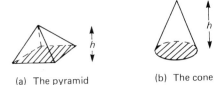

(a) The pyramid   (b) The cone

Figure 11.3

11.1.2   The prismoid

The prismoid is the commonest shape in earthworks calculations. A *prismoid* is a solid figure having two parallel end faces, not necessarily of the same shape, the sides of the solid being formed by continuous straight lines running from face to face. A prismoid may consist of any of the simple solids above, or it may be a truncated prism or wedge.

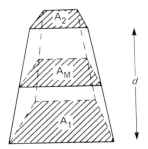

Figure 11.4

The common formula for all forms of prismoid is,

  Volume $= (d/6)(A1 + 4AM + A2)$ (11.4)

where $A1$ and $A2$ are the end face areas, AM is the area of the cross-section midway between the faces, and d is the distance between the end faces, as in *Figure 11.4*. AM may be obtained from either the calculated or measured side lengths.

11.2   Volumes from cross-sections

In roadworks, railways, canals and similar long earthworks, cross-sections are taken as described in Chapter 9, at regular intervals, then the volumes of cut or fill are obtained from these together with the measured distance between cross-sections. The solid figure between successive cross-sections is typically a prismoid, except where the longitudinal section is along the line of a curve or in special circumstances such as an opening in an embankment.

Three basic calculation rules are used, these being the *mean areas rule*, the *trapezoidal rule* and the *prismoidal rule*, all very similar to the ordinate rules in area calculations. In all the rules the individual cross-section areas may be obtained by any of the methods given in Chapter 10, but further suggestions on area determination are given in Section 11.2.6.

11.2.1 Mean areas rule

Where successive cross-section areas are $A1, A2, A3, \ldots An$, and distance from section $A1$ to section An is L,

$$\text{Volume} = (L/n)(A1 + A2 + A3 + \ldots An) \qquad (11.5)$$

In effect, the mean or average of all the cross-section areas is obtained and multiplied by the overall length of the excavation. The method is easy to apply, but not very accurate, giving too large a volume. It is therefore rarely used.

11.2.2 Trapezoidal rule

This method is also known as the *Method of end areas* or *Averaging end areas*. Where two successive cross-sections are $A1$ and $A2$ and they are spaced a distance l apart, the volume contained between them is

$$\text{Volume} = (l/2)(A1 + A2) \qquad (11.6)$$

provided the cross-section area at the midway point between $A1$ and $A2$ is actually the mean of the areas of these two. If there is only a slight change between successive cross-sections, the accuracy is acceptable for many purposes and the form may be developed for a large number of consecutive cross-sections $A1, A2, A3, \ldots, An$.

In this case,

$$\text{Volume} = (l/2)(A1 + 2A2 + 2A3 + 2A4 + \ldots + 2A(n-1) + An) \quad (11.7)$$

where l is the uniform distance between successive cross-sections. The similarity to the trapezoidal rule for areas will be evident. Again, the results are generally greater than the true volume.

11.2.3 Prismoidal rule

The best results are obtained if the volume of earth between successive cross-sections is a prismoid. The volume of the prismoid is obtained from Equation (11.4). Where a large number of cross-sections have been taken, every alternate section may be regarded as being an end face of a prismoid, the other cross-sections being taken as the respective AM or middle cross-sections of the prismoids. Then,

$$\text{Volume} = (l/3)(A1 + 4A2 + 2A3 + 4A4 + 2A5 + \ldots +$$
$$2A(n-2) + 4A(n-1) + An) \qquad (11.8)$$

where l is the uniform distance between successive cross-sections. Note that the value d in Equation (11.4) is equivalent to $2 \times l$ in Equation (11.8), hence $d/6$ becomes $l/3$. This is *Simpson's Rule* for *volumes*, similar to that for areas. Note that each cross-section appears once and their multipliers are respectively 1, 4, 2, 4, 2, 4, . . . 2, 4, 1, and there must be an odd number of cross-sections just as there is an odd number of offsets or ordinates in the area rule.

11.2.4   Prismoidal correction

During the railway building era it was common practice to use the Trapezoidal rule, for reasons of simplicity, rather than the somewhat more accurate Prismoidal rule. Calculation was tedious, of course, with no calculators, and much calculation was done with tables, graphs and nomograms.

If the Trapezoidal rule was used, a *Prismoidal correction* could be applied which would give the equivalent prismoid volume between successive cross-sections. The end result was better than that obtained by the Prismoidal rule, since it made use of the unmeasured middle cross-sections. This correction, however, required that the formation width remained constant within the prismoid and this was usually the case.

Since the introduction of calculating machines, the Prismoidal correction is little used, but it is

$$C = (l/12)(w0 - w1)(h1 - h0) \qquad (11.9)$$

where l is the distance between successive cross-sections, $h0$ and $h1$ are the depths of cutting at the centre-line and $w0$ and $w1$ are the horizontal distances between the tops of the embankments (see *Figure 11.5*).

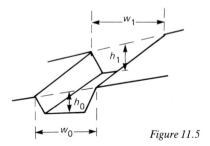

Figure 11.5

11.2.5   Roads and similar works curved in plan

Where roads, railways, etc., follow a curved line on plan, the volumes given by the standard formulae are incorrect, since the end faces are not parallel. The error is generally small and frequently disregarded for practical purposes. Details may be obtained in a specialist textbook on engineering survey.

11.2.6   Cross-section areas

Traditionally, cross-sections for volume calculations are plotted to scale, then the cut and fill areas deduced by planimeter, by counting squares, or by dividing the area into triangles and computing the triangle areas from scaled measurements. There are also formula methods and these may be more popular with increased use of calculators and computers. It should be noted, however, that where a cross-section is very complex, the data entry for a formula method by calculator may take longer than the traditional methods.

Formula methods rely, effectively, on making use of the reduced level at each change of slope of the cross-section, both of the existing ground surface and the final section shape, together with the distances of such points left or right of the centre-line. Most of these points may be determined, but the points where the new ground slopes intersect the existing ground slopes may have to be scaled or computed. In addition, the ground surface may vary irregularly across the width of the cross-section and the formation may not be horizontal. For roads, for example, the formation may be cambered, or superelevated. Further, many formula methods pose problems on cross-sections which are part in cut and part in fill.

If complex cross-sections are to be tackled by formula methods, then it is probably best to treat the levels and distances as co-ordinates and adapt the methods of Section 10.4. If a program is to be used, then the *Quick co-ordinate formula*, Equation (10.13), is generally suitable. The co-ordinates of the slope (new and existing surfaces) intersections must be obtained, either by scaling or using the following intersection formula:

$$RL_P = \frac{RL_F \cdot s - RL_G \cdot g - OD_F + OD_G}{s - g} \tag{11.10}$$

and

$$OD_P = OD_G + g(RL_P - RL_G) \tag{11.11}$$

where RL and OD are the reduced level and the offset distance from the centre-line respectively, s and g the side slope and the existing ground gradients respectively, (horizontal/vertical) and F and G are the points of known co-ordinates (offset and reduced level), as in *Figure 11.6*.

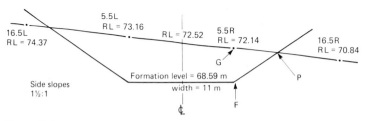

Figure 11.6

This method is suited to cuttings or embankments and offset distances right or left of the centre-line, but care must be taken over the sign of gradients. Two options for gradients are:

(1) If a gradient rises from a known co-ordinated point to an unknown, treat it as positive. If it falls, negative. Take all offset distances, whether left or right, as positive co-ordinates.

(2) Take the centre-line as the 'y' or vertical axis and the height datum as the 'x' or horizontal axis. Then as in normal co-ordinates, gradients in the first and third quadrants are positive, in the second and fourth quadrants negative.

From the example in *Figure 11.6*, the co-ordinates of P are:

$$RL_P = \frac{(68.59 \times 1.5) - \left(72.14 \times \dfrac{5.5 - 16.5}{72.14 - 70.84}\right) - 5.5 + 5.5}{1.5 - \left(\dfrac{5.5 - 16.5}{72.14 - 70.84}\right)} = 71.61\,\text{m}$$

and

$$OD_P = 5.5 + \left(\frac{5.5 - 16.5}{72.14 - 70.84}\right)\left(71.61 - 72.14\right) = 10.02\,\text{m R}$$

The corresponding point P on the left by option (2) is

$$RL_{P(\text{left})} =$$

$$\frac{(68.59 \times [-1.5]) - \left(73.16 \times \dfrac{[-5.5] - [-16.5]}{73.16 - 74.37}\right) - [-5.05] + [-5.05]}{[-1.5] - \left(\dfrac{[-5.5] - [-16.5]}{73.16 - 74.37}\right)}$$

$$= 74.06\,\text{m}$$

and

$$OD_{P(\text{left})} = [-5.5] + \left(\frac{[-5.5] - [-16.5]}{73.16 - 74.37}\right)\left(74.06 - 73.16\right) = -13.71$$
$$\text{or } 13.71\,\text{mL}$$

It may well, of course, be simpler to scale the co-ordinates of the intersection points if the cross-section has been drawn out.

The area of the cross-section by the quick method is therefore:

$$2A = x_F \cdot y_P + x_P \cdot y_G + x_G \cdot y_{cL} + \ldots - y_F x_P - y_P x_G - \ldots$$
$$= (5.5 \times 71.61) + (10.02 \times 72.14) + (5.5 \times 72.52) + (0 \times 73.16) +$$
$$(-5.5 \times 74.06) + (-13.71 \times 68.59) + (-5.5 \times 68.59) -$$
$$(68.59 \times 10.02) - (71.61 \times 5.5) - (72.14 \times 0) - (72.52 \times -5.5) -$$
$$(73.16 \times -13.71) - (74.06 \times -5.5) - (68.59 \times 5.5)$$
$$= 141.46 \text{ or } A = 70.73\,\text{m}^3$$

11.2.7   Changes in soil volume

Excavated materials generally expand and their volume becomes greater than when measured in the ground. This affects haulage costs, of course and may be an important consideration. The change in volume is known as *swell* or *bulking* and may vary from 5% to 50% of the original volume. Loose material, when compacted in embankments or fill, may conversely *shrink*. Sand may have 8% *shrinkage*, while a clay might be subject to 25% shrinkage. A soil survey may be needed to determine appropriate values for the area concerned.

11.3   Volumes from contours

Contour plans may be used to calculate volumes by treating the areas
enclosed by successive contours as 'cross-sections' and the vertical interval
between contours as the constant distance between cross-sections. The
three methods – mean areas, trapezoidal and prismoidal, may be used as
appropriate to calculate the volume of material contained between two
specified closed contours. This may also be applied to the calculation of the
amount of material in a stack such as dumped fill, sand-dunes and so on,
but the commonest application is probably the determination of the
volume of water which will be contained by a reservoir or dam. The
method may also be used to calculate the cut or fill in constructing a
horizontal or sloping base on an existing surface, as for the foundations of
an industrial site or the playing areas of a sports complex.

For water volumes, each contour line on the side slopes may be regarded
as one of successive water-lines and the successive plan areas of water at
the various levels are used.

Plan areas enclosed by a contour line are best measured by planimeter,
which is ideal for such irregular outlines. In calculating the volume of a
mound or hollow it is unlikely that the highest or lowest point will coincide
with the level of a contour line. It is therefore normal practice to calculate
the volume of the solid between this spot height and its adjacent contour by
the most relevant formula, i.e. $V = A \times h/3$, Equation (11.3).

11.4   Volumes from spot heights

The following method is most useful when an excavation is to be made with
vertical sides, such as for a basement. Spot heights are taken over the
whole area, on a uniform grid of squares, as outlined earlier for
contouring. Each square of the grid may then be considered as the top
end-face of a vertical prism running from formation level up to the original
ground surface. The volume of each of these regular prisms is the product
of the grid square area and the mean of the four corner heights. The corner
heights are, of course, ground level minus formation level.

Figure 11.7 represents the plan of an area to be excavated, with A, B, C,
. . . J being points on the surface which have been levelled in the form of a

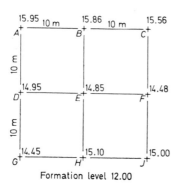

Formation level 12.00   *Figure 11.7*

square grid. The reduced levels are shown beside each point, the grid squares are 10 m on a side and the required formation level is 12.000 m above datum.

To calculate the excavation volume, each square prism could be treated separately and for each one the mean of the four heights taken and multiplied by the plan area of $10 \times 10\, \text{m}^2$. This would be a tedious procedure and unnecessary, since several of the corner heights appear in more than one prism. The calculation is normally made as follows, in tabular form:

Ground point	Pism corner height (h_n)	No. of squares occurs in (n)	Product $h_n \times n$
A	3.95	1	3.95
B	3.86	2	7.72
C	3.56	1	3.56
D	2.95	2	5.90
E	2.85	4	11.40
F	2.48	2	4.96
G	2.45	1	2.45
H	3.10	2	6.20
J	3.00	1	3.00
			Total = $49.14 = \Sigma(h_n \times n)$

Volume = $10 \times 10 \times 49.14/4$
       = $1228.5\, \text{m}^3$

In this calculation, each corner height is multiplied by the number of times it appears as a corner height and the total of 49.14 is the final sum of the four heights for all the squares. When this figure is divided by four, the result is the sum of the mean heights and need only be multiplied by one square area.

The mean height of the site, or in this example, the mean height of the excavation, may be obtained from $\Sigma(hn \times n)/\Sigma n$, which is 49.14/16 = 3.071 m (3.07125). The mean height $\times$ total area would also give the volume, in this example $3.07125 \times 20 \times 20 = 1228.5\, \text{m}^3$.

The grid could also be interpreted as a series of triangles, such as ABD, DBE, BCE, ECF and so on. The calculation would be similar, but each corner height would appear a different number of times and the quantity $\Sigma(h_n \times n)$ would have to be divided by 3 instead of 4 before multiplying by the plan area of one triangle. There is no advantage, however, in using the triangle method, unless the diagonal of any square lies along a ridge or a valley. If this happens, it may be better to divide the square into triangles, thus better emulating the surface of the ground. The calculation is, of course, more complex if there is a mix of squares and triangles.

11.5   Microcomputer applications

MICROSURVEY programs may be used for a variety of volumes applications.

11.5.1   Standard solids

The Trigonometry program includes routines for most of the common solids, together with the prismoidal formula and Simpson's rule for volumes.

11.5.2   Road earthworks volumes

The Road Volumes program accepts details of successive cross-sections, then computes the cumulative cut, fill and net excavation volumes at each cross-section chainage. At each chainage the user may enter either the reduced levels and distances for a two-level section, or the previously calculated cut and fill areas for the cross-section.

All data entered are stored for future reference and for this reason, memory limitations dictate the use of a two-level section only. More complex cross-section shapes may be plotted and measured by planimeter for cut and fill areas, or alternatively, they may be calculated first by the Trigonometry program and then the cut/fill areas entered into Road volumes. All the usual disk file data storage, amendment, printout and display facilities are provided.

11.5.3   Volumes from spot heights

Where spot heights have been taken on a grid of squares over a site, the volume above a specified formation level may be determined using the Grid Volumes program. Up to 200 points may be included in a set and any number of sets entered, since each set's data and results file is appended to the previous disk file. This has all the usual disk file data storage, amendment, printout and display facilities.

The printouts for the volumes programs are lengthy and are not illustrated here.

Part 4

Modern survey methods

Chapter 12

Bearing and distance methods

12.1 Introduction

Up to this point the problems involved in the survey tasks of supplying detail, height and control have been considered only in terms of the conventional and traditional methods of chain survey for detail, ordinary levelling for heights and the theodolite and band for control.

A variety of alternatives, however, are available, including the use of *bearing and distance methods, photogrammetry* and *map revision techniques*. Some of these methods use very sophisticated and expensive equipment and hence have been used mainly by the large land survey organizations. The pace of development of equipment has been so great over recent years, however, that many modern instruments are easy to use and may be hired for any desired period of time. Thus the individual who carries out surveys from time to time may have the choice of calling in professionals or of hiring equipment, the deciding factor often being which is the more economical alternative.

This chapter deals with the principles of bearing and distance methods, also known as *radiation* or *polar techniques*, using indirect forms of linear measurement. The more sophisticated types of equipment allow the supply of detail, height and control measurements to be carried out simultaneously. The equipment is expensive, but the additional costs may be offset by savings in time and man-hours. In the most advanced forms, survey data may be recorded magnetically in the field then transferred by modem and telephone line directly to a computer for automatic reduction and/or plotting. The cost of such facilities, however, is likely to be prohibitive to the 'occasional' surveyor and will not be considered here.

Indirect linear measurements may be made by *optical distance measuring equipment (ODM)* or by the more modern *electromagnetic distance measuring equipment (EDM)*. Optical distance measurement may be carried out using a theodolite, a level, or a special-purpose instrument such as a *tacheometer*. EDM can be carried out with an attachment to a theodolite, or using a special-purpose instrument resembling a theodolite.

12.2   Optical distance measurement

All optical methods of distance measurement depend on the relationship between an angle in a triangle and a distance, the α and *l* respectively in *Figure 12.1*. The angle may be fixed and the distance variable or alternatively the distance may be fixed and the angle variable. The application of the former method (fixed angle) is commonly termed *tacheometry*, although to some authorities the word covers all forms of optical distance measurement. Tacheometry is from the Greek *tachus* (swift) and *metron* (measure). An early form of English spelling was tachymeter and tachymetry, now lapsed.

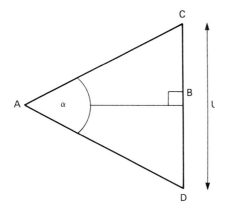

Figure 12.1

Tacheometry in its many forms is obsolescent, being rapidly superseded by EDM. There are still, however, many useful applications for the method.

12.2.1   Fixed distance/variable angle methods

12.2.1.1   *Subtense measurement*

This method is used to measure a single length, such as an inaccessible distance or a base line, but it has been used in traversing. The principle is illustrated in *Figure 12.1*, where AB represents the unknown length, *l* represents a known length established at right angles to AB such that CB = BD, then:

$$\tan(\alpha/2) = BC/AB,$$

or

$$AB = \cot(\alpha/2) \times l/2$$

If $l = 2\,\text{m}$, then

$$AB = \cot(\alpha/2) \text{ metres} \tag{12.1}$$

Instrument manufacturers produce two-metre horizontal staves of invar, known as *subtense bars*, to be used in conjunction with one-second theodolites. The subtense bar is obsolete for conventional survey, although it is still advertised by the makers as an accessory for industrial use in conjunction with an electronic theodolite and a microcomputer for accurate dimensional control to hundredths of a millimetre.

A variation of the subtense bar is the *sub-base*, in which the length *l* in *Figure 12.1* is of much greater length, and although CD should be at right angles to AB, CB need not be equal to BD. This is illustrated in *Figure 12.2*.

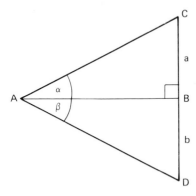

Figure 12.2

If EDM equipment is not available, this technique may be used to find the length of a base line of up to 1 or 2 kilometres in length over rough or undulating terrain, to an accuracy of approximately 1 part in 10000.

Referring to *Figure 12.2*, a base AB is selected such that a straight line CBD, the sub-base, may be established at right-angles to the base line. The points C, D and B must be visible from A. CB need not be equal to BD, but each must be at least one-twentieth of the length of the base AB. CB and BD are traditionally measured with a steel band, correcting for standardization, slope, temperatue, etc., to obtain their true horizontal distances.

The right-angles at B are set out with a theodolite, then the horizontal angles α and β must be measured correct to within one second of arc. Using a one-second optical reading theodolite, the angles would have to be read at least eight times on one face only, then

$$AB = CB/\tan \alpha = BD/\tan\beta \qquad (12.2)$$

and the two values to agree to at least 1 part in 10000. Note that no correction for slope is required on the line AB, since the theodolite at A records the angles in the horizontal plane, whatever the slope from A to B.

12.2.1.2 Rangefinders

The principle of these instruments is similar to that of subtense measurement, but the fixed base length is built into the instrument. Thus a fixed base is used and the base angles of the triangle are altered to bring two images of a distant target into coincidence, the distance to the target being read off a dial or scale.

Most rangefinders give slope distances. Their advantage is that the operator need not visit the point being fixed, the point may well be inaccessible. *Surveying rangefinders* do not appear to have been commercially viable propositions, and they may be considered obsolete. Commercial rangefinders available at present do not have the accuracy demanded by survey. Some rangefinders have a variable base principle, see Section 12.2.2.4

12.2.1.3   Tangential tacheometry

In this simple method, a theodolite is sighted on a distant vertical staff and the vertical angles to two points on the staff observed. If the vertical angles are θ and ϕ, and the distance between the two points on the staff is s, then horizontal distance is

$$d = s/(\tan \theta - \tan \phi) \tag{12.3}$$

as illustrated in *Figure 12.3*.

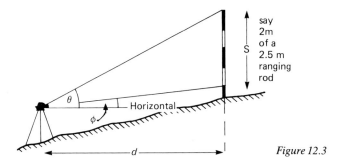

Figure 12.3

The method cannot be recommended, since even for plotted detail the accuracy is inadequate except at very short ranges.

12.2.2   Fixed angle/variable distance methods

12.2.2.1   Vertical staff stadia measurement with theodolite or level

Generally known as *tacheometry*, or *'tachy'*, the simplest form of this method uses an ordinary theodolite or surveyor's level with horizontal stadia hairs on the reticule, in conjunction with a levelling staff. It is a common method of providing detail and contours for 1:2500 or 1:5000 scale mapping. At the larger scales, the accuracy is adequate only over short distances, the linear distance accuracy being of the order of 1 in 400. This accuracy limits its usefulness for the provision of control for the mapping of a large area.

Tacheometric traverses using stadia measurement should generally be limited to a maximum of five legs in mapping work. The method has been used successfully for 1:25 000 scale mapping using a modified form of staff and instrument reticule.

In the chapter on levelling it was shown how the stadia hairs on the level diaphragm are used to check the centre hair readings and also the distance from the instrument to the staff. Thus, with a horizontal sight line and a vertical staff,

Horizontal distance = 100 × staff intercept,

or

$$D = 100\,s,$$

or

$$D = sk \qquad\qquad (12.4)$$

where k is the instrument multiplication constant, generally 100 today, s is the staff intercept and D is the distance from the centre of the instrument to the face of the staff.

Sixty years ago most theodolites had an additive as well as a multiplication constant and the formula was written

$$D = sk_1 + k_2$$

k_1 and k_2 being the multiplication and addition constants, respectively. Some modern telescopes have an additive constant, but it is normally so small with respect to the accuracy of stadia measurement that it is ignored.

(a) Horizontal telescope, $D = d = 100\ s$

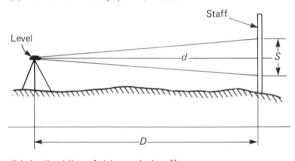

(b) Inclined line of sight-vertical staff

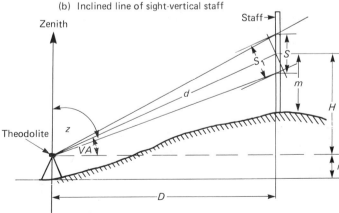

Figure 12.4

In addition, in older instruments the multiplier was often not exactly 100, but this problem no longer arises unless a reticule has been replaced in a telescope.

In stadia surveys using a theodolite, the line of sight is seldom horizontal, the typical case being rather as in *Figure 12.4(b)*. This shows a theodolite aimed on a staff held at a point which is higher than the instrument station, but it may be re-drawn to cover the case where the telescope is depressed. In the case illustrated, s is the staff intercept, but the inclined distance d is not equal to $100s$. In fact the inclined distance is equal to $100s_1$, and it is necessary to calculate s_1.

It will be observed from the figure that the angle between the faces of the real and imaginary staves s and s_1 is the reduced vertical angle, VA. The two small triangles formed by s, s_1 and the lines of sight through the stadia hairs are roughly equal, the larger angle in each triangle approximating to a right-angle, one being slightly greater and the other being slightly less than 90°. Taking these as being right-angles, then for practical purposes,

$s_1/2 = \cos VA \times s/2$, or $s_1 = s \cos VA$

From *Figure 12.4* it will be seen that for inclined sights, horizontal distance $D = d \cos VA$, but

$$d = 100 s_1$$
$$= 100 s \cos VA \tag{12.5}$$

therefore

$$D = 100 s \cos^2 VA$$

or

$$D = sk \cos^2 VA \tag{12.6}$$

If the zenith distance angle z is used, the expression becomes

$$D = sk \sin^2 z \tag{12.7}$$

For inclined sights, vertical distance

$$H = d \sin VA$$

therefore,

$$H = 100 s \sin VA \cos VA$$

or

$$H = sk \sin VA \cos VA \tag{12.8}$$

If the zenith distance angle z is used, the expression becomes

$$H = sk \sin z \cos z \tag{12.9}$$

If both D and H are to be calculated, then

$$H = D \tan VA \tag{12.10}$$

or

$$H = D/\tan z \tag{12.11}$$

The value H, of course, is the difference in height between the transit axis of the instrument and the centre hair reading on the staff, while what is required is the height difference between the reduced level of the instrument station and the ground point at the staff. Taking the instrument transit axis height above the station marker as i, and the centre hair reading as m, then in *Figure 12.4(b)* the required height difference is $i + H - m$, or in a more general form, $i \pm H - m$.

Thus, if the instrument station reduced level is RL_{inst}, the reduced level of the staff position is

$$RL_{inst} + i \pm H - m \qquad (12.12)$$

Equations (12.6) and (12.8), or (12.7) and (12.9) were used to deduce the required values for distance and height, but the work was laborious before electronic calculators were developed. Other forms of the equations may be developed, slightly easier to calculate, but reduction was commonly carried out by the use of special tacheometric tables and/or tacheometric slide rules or diagrams.

Throughout the long history of tacheometry, which is known to have been used by James Watt in 1771, manufacturers have developed a wide range of special instruments or attachments to help eliminate the calculation problems. The most successful of these special instruments have been the *diagram tacheometers*. In these instruments, the stadia hairs are actually lines on a diagram which change their distance apart as the telescope is tilted up or down, thus automatically correcting for slope. The values D and H are then obtained directly from the staff readings, with only the need for mental multiplication by factors such as 10, 20, etc., and these are often termed *direct reading tacheometers* (see *Figure 12.5*).

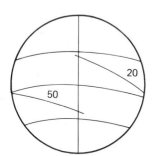

Figure 12.5

In some instruments the reticule carries a vertical and a horizontal line, but the sight line also passed through a second diaphragm, carrying the stadia lines, the second diaphragm being raised or lowered by a cam controlled by the vertical rotation of the telescope. In others, the diagram is etched on the glass plate of the vertical circle, or on a glass plate parallel to the circle, the sight line being deviated by prisms. The stadia lines are a pair of curves, their separation changing as the telescope tilts. This is illustrated in *Figure 12.6*.

The direct reading tacheometers have been displaced by EDM equipment, which is now fast and simple to use. Although electronic equipment becomes steadily cheaper, it is likely that precision optical

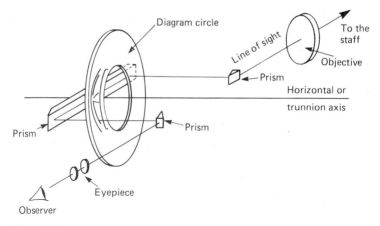

Figure 12.6

equipment will always be expensive, and few manufacturers still make these tacheometers. It should be noted, however, that direct reading tacheometers are still to be found in survey offices and instruments may be hired or reconditioned instruments purchased at reasonable cost. Since the accuracy of the direct reading tacheometers is similar to that obtained by using an ordinary theodolite for tacheometry, their use is limited to small areas of detail survey, and where control measurements are required it would be better to use EDM equipment.

12.2.2.2   Inclined staff stadia measurement with a theodolite

A variation on the stadia method is to hold the staff *normal* (at right angles) to the line of sight from the theodolite to the staff. This involves fitting the staff with a sighting device at the eye-height of the staff holder, the staff holder then inclining the staff forward until the sights are pointed on the centre of the theodolite. The reduction calculations are more complex, and there is little advantage in the method unless the angle of elevation or depression is steep. The method is little used and may be considered obsolete today.

12.2.2.3   Horizontal staff with wedge and parallel plate

If a glass wedge is placed horizontally in front of a theodolite objective lens, covering the middle third of the lens, two images will be visible when a target is viewed through the telescope. One image will be the view of the target brought by the light rays entering the telescope above and below the wedge. The other image will be the view of the target as seen through the wedge and this will be deflected to one side, thus two distinct images will be seen of the same target.

The refraction of light by the wedge and hence the deflection of the image, will depend upon the angle of the glass wedge. As illustrated in *Figure 12.7*, this is generally such that the distance from the wedge to the target is 100 times the deflection of the target image. In practice, when

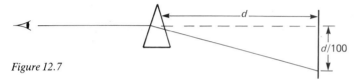

Figure 12.7

used with a graduated staff supported in a horizontal position, the direct image may be read at the vertical hair and the displacement of the second image may be read from the staff graduations. There are difficulties in reading the staff in this system, however, and generally a specially designed staff with an index mark is used, in conjunction with a parallel plate (as used in precise levelling) for fine reading. With the special staff and parallel plate, as shown in *Figure 12.8*, the index mark is made to coincide with a staff graduation and the straight line distance from instrument to staff (slope distance) may be read off directly if the staff is suitably graduated.

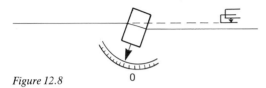

Figure 12.8   0

This principle is incorporated in a device (a *distance wedge*) which may be attached to the objective of a theodolite, with an appropriate horizontal staff supported on a tripod. Distances up to 150 or 170 m may be measured to an accuracy of 1:5000, and thus it is suitable for traversing as well as detail measurement; however, the equipment is obsolete now.

Optical distance measurement by distance wedge was of much higher accuracy than ordinary stadia measurement, but there still remained the need to calculate the reduced horizontal distances by multiplying the slope distance by the cosine of the vertical angle. The height difference also had to be calculated.

These problems were solved by the Swiss, Bosshardt, in 1923. He developed a theodolite-type instrument fitted with two wedges, each of half the refraction angle of the standard wedge, designed so that as the telescope tilted the wedges rotated in opposite directions, as illustrated in *Figure 12.9*.

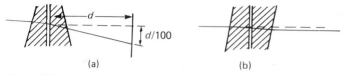

(a)   (b)

Figure 12.9

The wedge rotation causes a variation in the refraction of the light rays, such that the deflection observed on the staff gives the reduced (or 'direct') horizontal distance without calculation. This principle was incorporated in the Zeiss REDTA and later in the similar Wild RDH and the Kern DK–RT.

The DK–RT was taken up by the O.S., it was much used for surveys in built-up areas and it was also used to check motorway setting out. Easy to set up and use, capable of direct measurement of distances up to 166 m at about 1:10 000 accuracy and with a scale of tangents on the vertical circle so that $\triangle h = D \tan VA$, it is suitable for detail, heighting and traverse control. Although no longer made, instruments can be hired, or reconditioned instruments purchased, at about half the equivalent costs of the EDM equipment which is replacing it.

12.2.2.4   Fixed angle rangefinder

Figure 12.10 illustrates the principle of the *fixed angle rangefinder*. It incorporates a base rail, AB, and pentaprisms at A and B. The pentaprism angles are fixed, but one pentaprism can be slid along the base rail until the two prisms provide coincident images of the target. The triangle defined by the base rail and the two sight lines is some suitable proportion such that the target distance may be read off a scale which is engraved on the base rail.

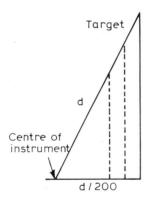

Figure 12.10

The commonest surveying instrument of this form is probably the Zeiss BRT–006 Self-Reducing Telemeter, introduced into the UK in 1962. This gives slope or horizontal distance direct, accuracy about 1:1500, range 60 m, or with a special staff, 180 m. It is simple to operate and is ideal for detail surveys at scales of 1:500 to 1:1250 if adequate control is available. The instrument is not in common use, having been introduced just before the developments in EDM.

12.3   Electromagnetic distance measurement

12.3.1   Background

Electromagnetic waves transmit energy through space and matter. Radio waves, infra red, visible light, ultra violet light, X-rays and gamma rays all travel at a speed of approximately 300 000 km/s in vacuum and slightly less in the atmosphere.

EDM in surveying may be considered to have had its origin in the 1939–45 war development of radio-location or radar. Assuming a fixed speed, a radio signal was 'bounced' off a target and the time interval between sending the signal and its return allowed calculation of the target distance. Although accurate enough for its purposes at the time, the method was not sufficiently accurate for survey work.

Research in Sweden and South Africa produced the first practical EDM instruments for surveyors. Dr Bergstrand of the Geographical Survey of Sweden, in collaboration with the company AGA, produced the AGA Geodimeter Model 1 instrument in the early 1950s. This was a long-range instrument for distances up to 30 km, using visible light. The instrument was placed at one end of the line to be measured and a mirror (*prism reflectors* today) placed at the other end, then the signal was transmitted along the line and reflected back to the instrument.

Dr T. L. Wadley of the South African Council for Scientific and Industrial Research at a similar time developed the Tellurometer MRA1, which used radio waves. Two transmitter/receiver instruments were used, a Master and a Remote, at either end of the line to be measured and equipped with radio communication links.

In each case the instruments gave a measure of the slope distance, which had then to be corrected for slope, temperature, barometric pressure, humidity and the curvature of the earth.

12.3.2   Measurement principle

Although visible light waves, radio waves, etc., all travel at the same speed, their wavelengths and frequencies may vary. *Wavelength* is quoted in *metres, frequency* in *cycles per second* or *Hertz* (1 cycle/s = 1 Hz), as illustrated in *Figure 12.11*.

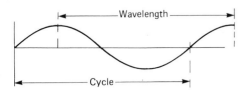

Figure 12.11

Taking the approximate speed of 300 000 km/s, a signal with a wavelength of 1500 m will have a frequency of

$$300\,000 \times 1000/1500 = 200\,000 \text{ cycles/s}$$
$$= 200\,\text{kHz}$$

VHF radio may use a frequency of 300 MHz, or 300 million cycles/s, giving a wavelength of 1 m.

Since it is impractical to measure the elapsed time between the transmission and return of a signal, modern EDM makes use of these wavelength and frequency characteristics by transmitting signals on more than one frequency, then determining the *phase difference* between the emitted and the returned signal at each frequency, as in *Figure 12.12*.

In practice, the EDM signal is a *carrier wave*, modulated to produce a measuring frequency and the instrument uses the phase difference of the measuring frequency to determine the slope or slant distance between the instrument and the retro-directional prism reflectors placed at the other end of the line being measured.

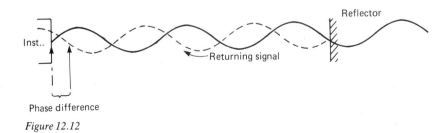

Figure 12.12

12.3.3   Development of modern equipment

The early instruments were medium to long range and would not have been suitable for the class of work covered by this text. By the mid 1960s, however, the traditional survey instrument makers were developing and producing EDM equipment and the early 1970s saw a new range of instruments coming into use. These instruments can measure distances from under 1 m up to 1, 2 or 3 km, typically to an accuracy of ±5 mm ±5 parts per million (usually written ppm), using modulated infra red light. The corrections for atmospheric conditions and altitude are switch-settable, and calculations which originally took much labour and an A4 page of computations are displayed directly by the instrument.

Today, then, a single electronic instrument can display slope and horizontal distances, horizontal and vertical angles and vertical height differences. Automatic data recording is possible and also the facility to carry out co-ordinate calculations in the field.

Chapter 13 illustrates the equipment which is considered, at present, to be economical for the survey tasks covered here. It must be appreciated, however, that new equipment is constantly appearing, that some models disappear as rapidly and there may be considerable fluctuations in costs.

12.4   Microcomputer applications

In bearing and distance survey for detail, possibly the only optical method of consequence today is vertical staff tacheometry. MICROSURVEY includes the Vtache program for vertical staff tachy and Edmtache, a program for EDM tachy detail survey reductions. It may be noted that Edmtache can be used for the reduction of horizontal staff tacheometry also.

12.4.1   Vertical staff tacheometry

The Vtache program may be used to compute the co-ordinates and reduced levels of detail points observed from survey stations using vertical staff tacheometry with a theodolite. Each observed point may be identified by a 5-character alphanumeric code, each station is identified by a 10-character name.

All the data and results from one instrument station are stored in a disk file, up to 100 detail points from a station. Any number of stations may be handled, since each station's data and results file is appended to the previous disk file. These disk files may be read back again, one by one, to produce displays or printouts of the results, or to allow the amendment of data.

Printouts may be obtained in full or abbreviated form. The latter provides only the information needed for plotting, i.e. point code, co-ordinates and reduced level. Full printout gives all data entered and all results.

12.4.2   EDM tacheometry

The Edmtache program is similar in operation to Vtache, but the instrument to staff distances observed by EDM are entered rather than staff readings and intercepts. The user may opt for entry of slope distances or horizontal distances, depending upon the equipment in use.

```
Leicester  Polytechnic

EDM  TACHE  —  DETAIL  PRINT

EDM Tache detail printout from CBM micro program EDMTACHE
─────────────────────────────────────────────────────────

     ■MICROSURVEY■ Software by Construction Measurement Systems Ltd
          Copyright (C) W.S.Whyte 1981/2/3/4  Tel: (0537 58) 283
          8000/2.1/8050/28 - for Leicester Polytechnic - 12.1.84
─────────────────────────────────────────────────────────

     Survey :  Leicester Royal Infirmary / Wild DI3

     Date   :  7th August 1984   (Data held on file edm-demo)

     Client :  R E Paul

     Surveyed by  HND Building Year 3 - 1979/80 on 9th May 1980

Tacheometer Station and Point Detail
─────────────────────────────────────────────────────────

Station   1 (401 - LRI )     Co-ordinates          Bearings

R.Level    60.100  Stn.       696.750   762.950 Observed  Computed
Inst.Ht     1.530  R.O.       794.000   591.900 236 07 00 150 22 47

Detail points from Station 1
─────────────────────────────────────────────

      H.Angle  Sl.Dist.  Sl.Angle  Tgt.Ht.       E          N      R.Level

      236 07 00  196.763  269 47 30   1.530    Point 1  (407ST)
Bg.   150 22 47  196.762 Dist. Computed values   793.999    591.901   59.385

      324 48 00   20.458  269 05 00   1.530    Point 2  (A    )
Bg.   239 03 47   20.455 Dist. Computed values   679.205    752.434   59.773

      327 13 30   42.223  269 06 00   1.530    Point 3  (B    )
Bg.   241 29 17   42.218 Dist. Computed values   659.652    742.798   59.437

      332 02 30   57.071  269 20 00   1.530    Point 4  (C    )
Bg.   246 18 17   57.067 Dist. Computed values   644.494     -
```

Figure 12.13

Figure 12.13 shows part of the full printout from a survey of part of the Leicester Royal Infirmary site, carried out using a Wild DI3 EDM attachment mounted on a Wild T16 theodolite. Vtache program printouts are similar.

12.4.3   Co-ordinate and trigonometry problems

Problems of linking to control (intersection, resection, etc.) may be handled using the co-ordinate and triangle solution routines in the Trigonometry program. The transformation of co-ordinates may be carried out with the Transform program and lists of co-ordinates may be sorted and printer listed with the Co-ordlister program.

Chapter 13

Bearing and distance equipment

13.1 Introduction

As stated earlier, most of the bearing and distance equipment can be used to supply the *detail, height* and *control measurements* from a single instrument set up at a station. This feature is its great advantage, since in these circumstances there are savings generally in both cost and time as compared with the more traditional methods.

The types of equipment illustrated in this chapter are those which are considered to be possibly the most economical in use at the present time, for the purposes of this text. Equipment with automatic data recording facilities are not covered fully, since the current prices are prohibitive unless the equipment is used to its full potential, i.e. used daily and taking full advantage of all the facilities available. Nevertheless, if time is critical and cost immaterial, the more sophisticated equipment will provide the fastest results.

Most users must pay attention to cost and for the occasional user the ideal instrument for any task is a simple EDM instrument attached to a theodolite. If the use is infrequent, say once or twice a year only, it may be cheaper to hire or lease the equipment. If the survey accuracy is such that stadia measurement is acceptable and the survey is small (no added control needed), then a theodolite and a levelling staff can be the cheapest solution.

The more frequently that surveys are to be carried out, then the better should be the equipment. Thus, if EDM surveys are carried out often, the instrument should be capable of giving horizontal distance direct and of displaying the co-ordinates of the surveyed point. If much setting out is to be done using EDM, the instrument should be able to indicate how much a tentative point needs to be moved away from, or towards, the instrument, or left or right. The additional costs will be offset by the savings made with the added facilities over the extended use time.

There may be occasions when it might be marginally better to use one of the self-reducing tacheometers or a fixed angle rangefinder, but these are likely to be rare. They are illustrated here for the benefit of individuals who may be unfamiliar with these but may wish to make use of them.

13.2   Optical distance measuring equipment

Figure 13.1 illustrates examples of the three optical distance measuring instruments referred to above. These are currently available as new, re-conditioned, or for hire.

The Wild RDS self-reducing diagram tacheometer (vertical staff) is illustrated in *Figure 13.1(a)*, together with a typical staff reading as viewed through the telescope. Similar in appearance to a theodolite, it may be used as a simple theodolite, giving angles by estimation to 6″ of arc. The accuracy of distance readings is a little better than that obtained with stadia measurement by theodolite or level.

The Kern DK–RT self-reducing tacheometer (horizontal staff) with rotating wedges and two parallel plates is illustrated in *Figure 13.1(b)*, together with a view of the horizontal staff as seen through the telescope. The addition of the micrometer drum reading gives the horizontal distance. The height difference is given by

$$\triangle h = \text{Horizontal distance} \times \tan VA,$$

the instrument including a scale of tangents of the vertical angles. The accuracy of a single distance reading is roughly 1:5000.

The Zeiss Jena BRT–006 Telemeter, a fixed angle rangefinder-type device, is illustrated in *Figure 13.1(c)*. The distance accuracy is approximately 40 mm up to a range of 60 m and about 1:1500 at greater distances.

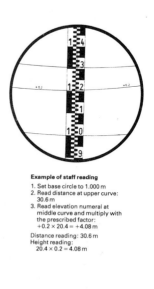

Example of staff reading

1. Set base circle to 1.000 m
2. Read distance at upper curve:
   30.6 m
3. Read elevation numeral at
   middle curve and multiply with
   the prescribed factor:
   +0.2 × 20.4 = +4.08 m

Distance reading: 30.6 m
Height reading:
   20.4 × 0.2 = 4.08 m

Figure 13.1a Wild RDS self-reducing diagram tacheometer

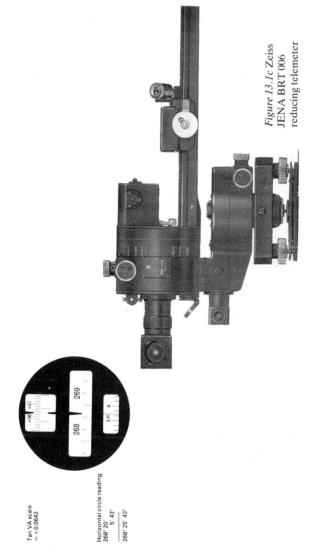

259

Figure 13.1c Zeiss
JENA BRT 006
reducing telemeter

Staff reading on main index
85 m (or 50 + 35)
0.275 m (from micrometer drum)

85.275 m

Tan VA scale
= + 0.0642

Horizontal circle reading
268° 20'
      5' 43"

268° 25' 43"

Figure 13.1b Kern DK-RT self-reducing
tacheometer

13.3 Electromagnetic distance measuring equipment

13.3.1 Simple EDM attachments

For the occasional user, the simplest and cheapest forms of EDM equipment, giving slope distance only, are ideal. Adaptors are available to fit these to the telescope of most theodolites. Corrections to the observed distance may be made in the field or the office, as preferred, using a pocket calculator.

The cost of this EDM equipment together with battery, battery charger, reflectors, reflector supports, telescope adaptor and possibly other accessories, is roughly comparable to the cost of a one-second theodolite. Note, however, that on most tasks a 20″ or 6″ theodolite will be adequate.

The following points should be considered when purchasing:

The accuracy. All makers of this basic form of EDM quote ±5 mm ±5 ppm.

Reflector prism ranges. With a single prism, manufacturers variously quote distances from 300 m to 1400 m. Additional prisms may be needed to reach the desired ranges, present cost of these is £150 to £250 each.

Operating temperature. Generally between −20°C and +50°C.

Interval between battery charges, number of batteries needed for a day's work. Manufacturers quote from 1 to 10 h on different models.

Maximum angle of elevation or depression. Limited to ±30° for some models, adequate for most tasks.

Servicing or replacement facilities in the country of operation.

Three simple forms of EDM equipment are illustrated in *Figure 13.2*, together with the manufacturer's brief specification. These are

(1) the Japanese Sokkisha REDmini,
(2) the Swedish AGA Geodimeter 110, and
(3) the Swiss Kern DM 102.

13.3.2 Sophisticated EDM attachments

More sophisticated EDM instruments for attaching to a theodolite are available, giving automatic display of horizontal distance and height difference. These typically cost some 60% more than the simple attachments and may not be worth the added cost for the infrequent user.

Two sophisticated EDM attachments are illustrated in *Figure 13.3*, these are

(1) the AGA Geodimeter 122, and
(2) the Sokkisha RED3.

The two models have many similarities, including accuracy, temperature range, automatic vertical angle sensor for horizontal distance and height difference display, LED digital displays and even price. Both have options to calculate the height of a remote object vertically above the reflector. Both have additional options, thus the Geodimeter 122 has an interface for a data logger and data transmission to a computer and a one-way communication system for use in conjunction with built-in visible light

Figure 13.2a   *Figure 13.2b*   *Figure 13.2c*

	Figure 13.2a	*Figure 13.2b*	*Figure 13.2c*
Accuracy	± 5 mm ± 5 ppm	± 5 mm ± 5 ppm	± 5 mm ± 5 ppm
Maximum range, one reflector (m)	300–500	1000	1000
Operating temperature (°C)	−10 to +50	−20 to +50	−20 to +50
Battery measuring time, fully charged (h)	1	–	10
Display mode	LCD	LED	LCD
Range of tilt	–	30°	45°
Price	Relatively cheap	–	Most expensive

beams for setting out, making it the ideal setting out instrument. The RED3 has an option to obtain, on site, the rectangular co-ordinates of the reflector position and when setting out the difference in length between the measured and the required horizontal distance.

If an instrument has a built-in calculator, then the maker may program a number of different options.

13.3.3   Independent EDM instruments

Where EDM survey is to be carried out on an average of at least one day per week, then it may be sensible to consider the independent EDM instruments. These are equivalent to electronic theodolites with built-in EDM and are often called electronic tacheometers, or electronic and modular total stations. They cost from two-and-a-half to four-and-a-half times as much as the simple EDM attachments, but may be worthwhile if much computation is involved and the use is frequent, e.g. in setting out by rectangular co-ordinates.

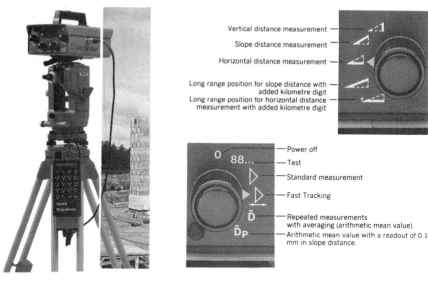

Vertical distance measurement

Slope distance measurement

Horizontal distance measurement

Long range position for slope distance with added kilometre digit
Long range position for horizontal distance measurement with added kilometre digit

Power off

Test

Standard measurement

Fast Tracking

Repeated measurements with averaging (arithmetic mean value).

Arithmetic mean value with a readout of 0.1 mm in slope distance.

Figure 13.3a

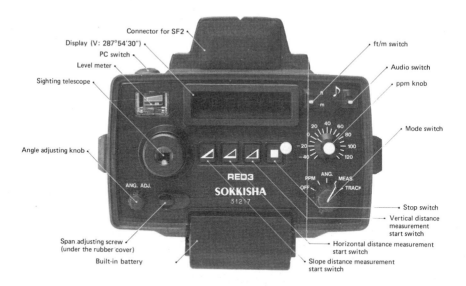

Connector for SF2

Display (V: 287°54'30")

PC switch

Level meter

Sighting telescope

Angle adjusting knob

ft/m switch

Audio switch

ppm knob

Mode switch

Stop switch

Vertical distance measurement start switch

Horizontal distance measurement start switch

Slope distance measurement start switch

Span adjusting screw (under the rubber cover)

Built-in battery

The function keyboard SF2

The function keyboard greatly extends measuring capabilities. In addition to slope, horizontal and vertical distance measurement (which is already a feature of the RED3), stake out measurement and X,Y-co-ordinate measurements come within range.

Figure 13.3b

Figures 13.4, 13.5 and *13.6* illustrate a range of equipment and complete surveying systems. The equipment illustrated and mentioned here is portrayed simply on the basis of up-to-date and suitable literature available from manufacturers and agents at the time of writing. The fact that a particular make or model has not been referred to should not be taken as indicating that it is in any way inferior to those shown.

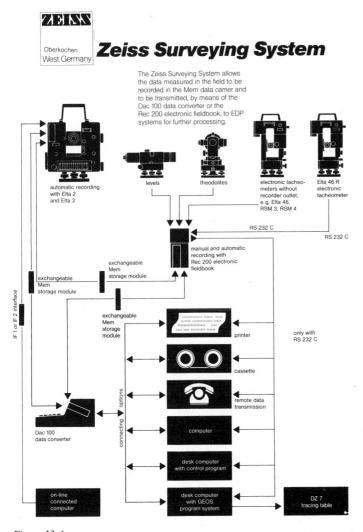

Figure 13.4

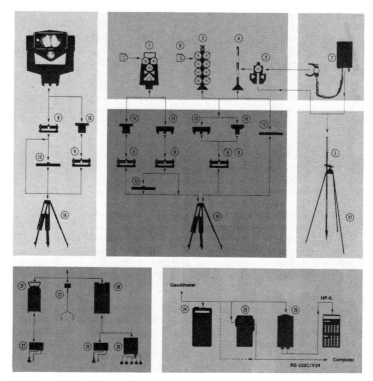

Geodimeter 140 is part of a complete system in which each individual part offers maximum flexibility. A system which has design features for many survey applications without sacrificing accuracy or performance in any one.

1. Traversing target for one to three prisms. 2. Tiltable reflector with target. 3. Prism assembly. 4. Foot and signal pin. 5. Range pole glassfibre. 6. Prism. 7. Unicom receiver. 8. Geodimeter tribrach (supplied with Geodimeter 140) or Wild type tribrach. 9. Zeiss type tribrach. 10. Adapter Geodimeter 140 to Zeiss tribrach. 11. Adapter traversing target to Zeiss tribrach. 12. Adapter Geodimeter tribrach to Kern tripod. 13. Tribrach adapter to Geodimeter and Wild types tribrach. 14. Tribrach adapter to Zeiss tribrach. 15. Tripod adapter to Kern tripod. 16. Tripod Wild, Zeiss or Kern. 17. Range pole support. 18. Battery 2 Ah (2 pcs is supplied with Geodimeter 140). 19. Charger for one battery, 220 V a.c. 20. Charger for four batteries, 220 V a.c. or charger for four batteries 115 V a.c. Heavy duty battery, 6 Ah. 22. Charger for heavy duty battery, 220 V a.c. or charger for heavy duty battery, 115 V a.c. 23. Adapter cable for 'car battery'. 24. Geodat 122. 25. Serial interface (RS-232C/V24). 26. HP-IL interface for HP41 or HP75 calculator.

Figure 13.5 Geodimeter 140 measuring system

EDM equipment available at the present time in the UK includes:
AGA Geodimeters
Kern DM models
Keuffel and Esser (USA) Auto Ranger
Nikon (Japan)
Omni 1 (USA)
Pentax (Japan)
Sokkisha (Japan)
Tellurometer
Topcon (Japan)
Wild range
Zeiss Jena (East Germany)
Zeiss Oberkochen (West Germany)

DM502 Electro-optical distance meter attachable to the theodolite telescope
Compact, lightweight, infrared distance meter, attachable to the telescopes of the theodolite series DKM2-A, K1-M and K1-S.

Reflectors
Trihedral prism reflector in sturdy plastic housing, serves also as target. Target height same as height of telescope axis.

R48 Recording unit
The R48 is used for storing data acquired in the field. The measured values can be entered manually or transmitted automatically form the electronic theodolite. Supplementary data can be entered via the keyboard. The data stored in the memory can be retrieved and verified at any time. It is not erased when the instrument is switched off and remains available for about a month. The storage capacity is 48 000 digits which suffices for recording up to 800 points in polar co-ordinates. Data transfer to a computer is via the V24 interface.

E1 Electronic theodolite
Compact and easy-to-operate electronic theodolite. Clear liquid crystal display in both telescope positions. Constant horizontal axis height. High accuracy of vertical angle measurement by liquid compensator. When the DM502 distance meter is mounted, the integral microprocessor automatically computes horizontal distance and elevation difference. Selector switch for angle display in gon or degrees and for distances in metres or feet. Index error display after sighting in position II. Illuminable reading windows.

RD10 Remote receiver RD10
The RD10 simplifies layout work by receiving the values transmitted from the distance meter and displaying them digitally at the reflector location. The display provides slope distance, horizontal distance, elevation difference horizontal direction, vertical angle longitudinal and transverse deviations, depending on the instruments used. Selector switch for display. Bayonet joint for convenient attachment to DM502 reflector.

Reflector   Remote Receiver

Elevation different – 8.043 m
Horizontal distance 98.486 m
Vertical angle 105.188 gon
Horizontal direction 130.120 gon

DIF41 Data interface
The DIF41 permits the automatic transmission of measurements from the DM502 electro-optical distance meter or from the E1 electronic theodolite to a Series 41HP calculator. The DIF41 can be used in combination with the DM502 electro-optical distance meter on K1S, K1-M, DKM 2-A and E1 theodolites. The programmed HP calculator processes these values and transmits the results to the RD10 remote receiver at the reflector location via the DM502. Data communications between the Hp calculator and a peripheral device are possible with a V24 (RS 232) interface. The DIF41 is also designed to house the HP calculator.

Figure 13.6 Kern modular instrument system

13.4 Maintenance

It must be emphasized that, as with all survey instruments, the equipment covered in this chapter must be regularly checked, tested, adjusted, maintained and, where appropriate, calibrated. All this must be done in accordance with the manufacturer's instructions and adjustments should generally be carried out by a skilled instrument mechanic.

Chapter 14

Bearing and distance surveys

14.1 Introduction

This chapter outlines recommended techniques for the use of short-range EDM equipment in large-scale topographical mapping. Minor modifications will be needed if optical distance measuring equipment is used. Similar techniques are employed in cadastral mapping, usually without heighting and also in hydrographic work. In the latter case, the target may be a moving boat, then more than one set of reflectors may be used, each set at a different angle but on the same support in order to ensure a return signal. The reverse process of *setting out*, or *radial stake out* as it is sometimes known, is again similar and is further dealt with in Chapter 17.

14.2 Basis of the method

Detail and spot heights are supplied by linear and angular measurements taken at an instrument station to all the points of detail and spot heights required to be surveyed. To do this, a reference direction is established at the station and the angle between the reference direction and the line to a detail point is measured, together with the distance from the instrument to the detail point, the procedure being repeated for all the detail points and spot heights.

The detail and spot heights are plotted by the use of a protractor and scale, or using a polar co-ordinatograph, or by converting the polar co-ordinates (bearing and distance) to rectangular co-ordinates and plotting these on a rectangular grid. If available, a rectangular co-ordinatograph may be used to plot rectangular co-ordinates, or they may be plotted by computer. Where the area to be surveyed is large, many instrument stations may be needed and these may be connected together by triangulation, trilateration, or, very often, traversing.

14.3 Advantages and disadvantages

The primary advantage of the method is that using one set of equipment, at one instrument station, all the detail, height and control measurements in the area may be observed and measured. The principal disadvantages of the method are the initial capital cost of the equipment and its bulk and weight. Each new model, however, seems to be lighter and smaller than its predecessor. The equipment can be operated by a minimum team of two, but a team of at least three is generally more economical.

On a survey which only requires to show detail at a scale of 1:500 or smaller, it is generally preferable to simply use chain survey methods, unless there are obstacles to chaining and/or traverse control is demanded. Many large-scale surveys are carried out for 1:200 scale plans, but at this scale chain survey is not generally of sufficient accuracy unless the lines are laid out using a theodolite, or a level, etc. Bearing and distance techniques are recommended at this scale. Where a site survey requires spot heights and/or contours to be shown, it will normally take fewer man hours to carry out the task by bearing and distance methods. Similarly, if traverse control is required with or without heighting, it should take less man hours and be cheaper in cost by bearing and distance methods, even if the team spend only three months of each year on survey.

14.4   Equipment

The following list covers a sample set of equipment for a task including traverse control and using a simple EDM attachment.

Theodolite plus adaptor for the EDM instrument.

EDM distance meter (or *measuring head* or *aiming head*) and control unit.

Battery, plus spare battery.

Charging unit (may or may not be built into the battery). Check necessity for adaptor plugs for mains supply.

Battery cables.

Reflectors, prisms with targets (2).

Tripods (3).

Prism poles (2). Some models telescopic, some with extension pieces, sometimes ranging rod may be used.

Holders, tribrachs, adaptors to fit prisms and targets to tripods and prism poles.

Station marking equipment.

Data recording equipment.

Plotting equipment, including a 360° protractor or polar co-ordinatograph. The protractor radius should be at least as great as the maximum measured length to be plotted at the plotting scale. (It is always bad practice to extrapolate beyond the circumference of a protractor. The alternative is to convert to rectangular co-ordinates.)

14.5   Preliminary considerations

A number of considerations must be settled before the fieldwork can be commenced.

14.5.1   Suitability of bearing and distance methods

It may be that other methods are more suited to parts of the task. On occasion it may be found that chain survey and/or map revision techniques are suited to part of a survey which is principally being carried out by bearing and distance methods. See Chapter 4 and Chapter 16 for details. Where building elevations are concerned, photogrammetric methods may be appropriate and similarly in quarries, etc. See Chapter 15.

14.5.2   The survey scale

The scale of the survey will affect the limits of the linear and angular measurement, the maximum length of traverse which may be used and the equipment required. If the scale has not already been agreed upon, then its choice will be affected by the following factors:

The amount of detail and other information to be shown.
The area to be covered.
The size of the completed plan or map and its convenience to the user.
The purpose of the plan or map.
The accuracies required.

14.5.3   The need for additional control

If a site is extensive, it may be necessary to provide additional control in order to achieve a desired accuracy. Additional control measurements may be needed to tie the survey to national or municipal control. As a general guide, it is recommended that an EDM traverse for 1:500 scale mapping should not exceed 20 legs or 2000 m in length and proportionally at the other scales in use. For example, 1:200 scale, 8 legs and 800 m in length, but 1:1000 scale mapping, 40 legs and 4000 m in length.

14.5.4   Sight length limitations

The sight length available will affect the layout of stations and of traverses. Generally the EDM equipment itself does not impose limitations, rather it is the survey scale and the atmospheric and environmental conditions. Signalling arrangements are important and visibility, affected by the presence of mist, smoke, haze, etc. The angular accuracy of the theodolite may impose a limitation, thus an arc of 10 mm is subtended by an angle of $20''$ at a distance of 100 m. In built-up areas, stations may need to be within 60 to 80 m apart to ensure that all detail can be surveyed accurately.

A limitation may be imposed by the use of a protractor for plotting, thus a 300 mm diameter protractor with a plotting scale of 1:500 limits the sighting distance to 500×150 mm $= 75$ m.

14.6   Field procedure

On arrival on site and before making any observations, the surveyor should carry out the usual thorough reconnaissance of the survey area. He should make himself familiar with the general layout and, in particular, identify any likely problems.

14.6.1   Instrument stations

14.6.1.1   *Selection and marking*

Any instrument station should provide a clear view of the area to be surveyed from the station, bearing in mind the limitations mentioned in Section 14.5.4. Where the ground is flat, station points should be selected on elevated positions if at all possible, e.g. low bare hills in open countryside, low flat roofs, raised platforms etc., in built-up areas. If the survey is to be controlled by more than two instrument stations then the stations must be tied together and this generally means that at least two other instrument stations must be visible from any instrument position.

The stations should normally be tied together by traversing, either a single loop traverse, or possibly a network of traverses. On a long narrow site they may be tied to a base line or to some other form of control such as the O.S. National Grid. Triangulation may also be considered as a possible solution on occasion. Traverse layout is covered in Chapter 6.

Stations may be marked with a nail, wooden peg, rivet, etc., depending upon the ground conditions, but it is preferable to use an existing man-made feature if possible. Such features include a point on a manhole cover, corner of a drain gulley grating, etc., these being longer lasting than artificial marks and easier to re-locate if observations must be repeated or revision is needed. If centring tripods are used, the station mark should be centre punched to facilitate the positioning of the centring rod.

14.6.1.2   *Sketches*

When the selection and positioning of a station is complete, it is traditional to draw a sketch of the area showing the detail to be surveyed. Normally a separate sketch is drawn for each station and in a large survey department these will be produced to a rigid set of rules similar to those referred to in chain survey, so that all members of the department will fully understand the content of the sketches. Smaller organizations tend not to be so rigid.

Where the survey area is small, a single sketch may be used for the whole site. If the detail to be portrayed is simple, e.g. rows of hedges or a series of spot heights, then the sketch may be dispensed with and a simple annotation placed in the booking sheet adjacent to the relevant field measurements, a sketch only being added for intricate detail. Any sketch produced must be clear and reasonably accurate, the objectives of the sketch being:

To make clear to the plotter how the surveyed detail is linked to the observations, how it is to be shown on the plan and to what other detail it is connected.

To indicate any taped supplementary measurements such as plus measurements required to complete the plotting of a rectangular feature.

To indicate any names and their positioning on the drawing.

To indicate the possibility of any gross errors.

To assist in the interpolation of contours.

To indicate to the reflector holder the position of the points of detail and the approximate positions of spot heights.

To ensure that no point to be surveyed is forgotten by the observing team.

To show the position of the instrument station and reference object in relation to the detail.

To show what detail, if any, has been surveyed from more than one instrument station.

To indicate detail points lying on the same alignment.

14.6.1.3   Coding systems

Despite the usefulness of the traditional sketch as described above, there is a trend today to dispense with the sketch and instead use a system of short standard alphanumeric codes recorded in the field sheet or recorded by electronic data loggers. Even though this requires assistants with higher skills than needed in chain survey, users of modern EDM are finding it faster and cheaper to use coding systems.

Where electronic tacheometers are used with data loggers, a coding system must be used for recording detail and although such systems are likely to be uneconomical for the occasional user at present, these costs are falling. It might, therefore, be advantageous to adopt such methods now for hand booking, so that there will be minimum upheaval when more sophisticated equipment comes into use. A similar statement could apply, of course, to the use of automatic plotters.

In the remainder of this chapter it will be assumed that data will be recorded by codes and that the traditional sketch will not be used.

14.6.1.4   Setting up the instrument

As with all specialist equipment, the instrument should be set up in accordance with the manufacturer's instructions. The accuracy of centring over the station mark required is dependent upon the accuracy demanded of the observations, whether they are only for the supply of detail and spot heights or for control observations for computation. If spot heights are to be observed, then the height of the centre of the instrument above the station mark must be measured and recorded. If a centring tripod is used, with central graduated plumbing rod, the instrument is readily set up at a convenient working height.

14.6.1.5   Selection of local reference object (LRO)

The *reference object (RO)* at a traverse station is usually the back object, but if a large number of detail points are to be recorded at a station it will be advantageous to select a *local reference object*. The LRO provides a rapid means of checking the orientation of the instrument at any time during the round of observations at the station. The LRO may be any well-defined object, a point of detail which must be surveyed, or even the RO proper. The latter, however, is often inconspicuous or unavailable once the tripod has been removed from the back station, hence the recommendation to use an LRO. If an LRO is selected, it should be observed at least at the commencement and at the end of the round of observations.

14.6.2   Observations

14.6.2.1   Control

The *control observations* required at a station should be made preferably before the observations to points of detail and spot heights, to avoid the danger that they may be forgotten. Control observations should be made in accordance with the recommendations of Chapter 6, but it should be noted that with some EDM equipment it is not possible to measure on both faces. An alternative is to read on the one face but on at least two zeros. The LRO, if any, should be observed at the same time as the control observations are made.

14.6.2.2   Detail and spot heights

Detail points and spot heights should be observed on completion of the control observations. They may be observed in any order, but clearly a systematic approach is desirable. It is expedient to observe the detail and spot heights in strings, a string being a series of related points such as the line of a hedge, or a kerb line, or a row of trees, or a building, etc.

With the cheapest forms of EDM equipment, one vertical angle, one slope distance and one horizontal angle to each point of detail or spot height is normally all that is necessary. The LRO should be re-observed on completion of the detail and spot height observations.

14.6.2.3   General

A survey team should consist of a minimum of two, the senior being the team leader responsible for the survey. The team leader decides which detail is to be surveyed and the positions for spot heights and holds the reflector, and also records all the relevant string information except the angles and distances taken at the instrument station. The assistant observes and records the angular and linear measurements taken at the instrument station to the reflector positions.

If separate booking sheets are used at a station, they should be fastened together at the end of the day's work. Alternatively, some surveyors use VHF radio telephone communication to transfer all information on to one booking sheet, or enter all data in a data logger. Again, data may be coded into an electronic tacheometer, along with the automatically recorded observations. Where there are three members in a team, the third member (field assistant or trainee surveyor) can position a second reflector under the direction of the team leader.

14.6.3   The reflector

14.6.3.1   How to position

The prism reflector must be placed at the point of detail such that its face is normal to the line of sight from the instrument. It may be tripod-mounted, or fixed to a prism pole, or even hand-held on occasion. Note that where a prism pole with circular bubble is used, keeping the bubble central is more important than accurate aiming of the reflector.

14.6.3.2   Where to position

For control observations the reflector will normally be placed on the back and forward stations and possibly on other station points. For these observations it should preferably be centred over the particular station mark on a tripod.

For detail and spot height observations, the reflector should be placed on or over the actual point. Sometimes this is impractical and then it should be placed, as convenient, either in front of or behind the point on the line from the instrument, the displacement being noted. Again, if this cannot be done, it may be held to the left or right of the point and the left or right offset noted.

14.6.4   Alternative measurement techniques

Sometimes detail may be inaccessible, or invisible from the instrument station and additional measurements may be needed which cannot be obtained by bearing and distance methods. The following alternative methods may be used, but their use should be limited in order to avoid confusion for the plotter:

Plus measurements to define rectangular detail.
Detail fixing by intersection.
Widths of parallel features.
Chain survey methods (Chapter 4).
Map revision techniques (Chapter 16).
The use of a cul-de-sac station.

A *cul-de-sac station* is one from which only one other instrument station is visible, sometimes known as the *parent station*. An example might be a station set up solely to measure the detail in an enclosed yard. The location fix of a cul-de-sac station is obviously weak, since it cannot be linked to other stations for a check, hence it should not be used unless there is no economical alternative. A possible checking procedure is to survey a common point of detail from both the parent and the cul-de-sac station, and co-ordinate plot the detail from both stations.

14.6.5   Communication between team leader and observer

A good communication system between these individuals is necessary in order that each may be aware of the other's requirements. If VHF radio telephones are not available then a simple hand signalling system can be effective. Over long distances, a white handkerchief or a coloured flag may be of assistance. Any such system must be simple, well understood and unambiguous.

The team leader needs to be able to indicate to the observer that the reflector is in position for observing. The observer must be able to indicate:

Measurements complete, move on.
Come to me.
Go back to previous point.
Reflector obscured.
Raise reflector.
Etc.

If separate booking sheets are being used, then it is advisable for the two to check with each other, say at every tenth reading and at the end of observations, to ensure that no reflector points have been missed and the reflector point numbers on both sheets are the same.

14.6.6   Recording observed data

The method of recording data recommended here assumes two booking sheets, one for use by the observer and the other for use by the team leader. If VHF radio is used, then the data may be combined on to a single sheet in the field. If data logging or automatic recording of the observations is used, then the team leader's data must be coded to suit the particular equipment in use. The coding shown in the example booking sheets may be suitable, some coding systems allowing numeric codes only and some accepting alphanumeric codes.

The essentials of all bookings are clarity and accuracy, so that the field bookings may be understood by all members of the team. Alphanumeric characters should be printed, corrections should be made by ruling a single line through the characters and printing the correct entries above or below the original. No overwriting should be permitted.

The booking sheets may show:

The name of the task and traverse.

Team names, leader, observer, etc.

Survey date, weather.

Identity of equipment in use (theodolite, EDM, reflectors, etc.) and any additive constants or similar values.

Station number, description, height, instrument height.

Control observations; stations observed, vertical angles, slope distances, horizontal angular readings.

Detail and spot height observations; point number, other points connected to the point, line or area symbol to be used by the plotter.

Also generally:

Symbol identification annotations.

Annotations to assist interpretation of contours (ridge or valley lines etc.).

Corrections to be applied when reflector is not on the point being surveyed.

Measurements and type, if any, for the alternative techniques set out in Section 14.6.4.

Figure 14.1 illustrates an observer's booking form, designed for an instrument in which the distances could be read only on face right. The upper part of the form is for station data common to all observations, the next part is for control observations and the lower part is for detail and spot height readings.

Figure 14.2 illustrates a form which may be used by the team leader. This indicates how detail is to be portrayed, how it is connected and the supplementary measurements. Note that the data will be coded and coding

Bearing and Distance Surveys by use of FDM

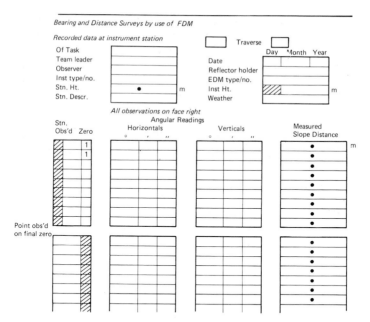

Figure 14.1

Bearing and Distance Surveys by use of EDM

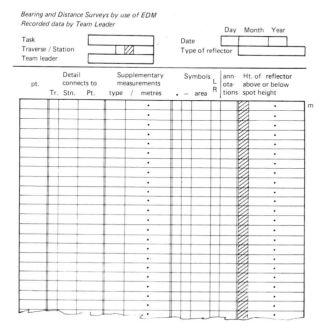

Figure 14.2

may vary from the very simple to the very comprehensive. Too basic a system may inhibit proper booking, while too complex a system may unduly slow down the work and some sensible compromise is desirable. The lack of sketches may mean that some minor items of detail are not fully annotated, but if the plotted plan is attached to a small drawing board, e.g. A2 size, then map revision techniques may be used to complete the drawings.

The upper part of the form is for data common to all detail and height points. The remainder of the form is divided into five vertical blocks. The left-hand block contains the detail and height point numbers, then all information on the same line across the sheet refers to the point number in the left-hand block.

The second vertical block is used to indicate what detail, if any, the point in the first block connects to. If the detail is to be observed in strings, then in many cases it will join back to the previous point of detail and the annotation 'pp' (for previous point) may be adequate. In other cases it may be necessary to indicate the possible traverse, station and point number, for example 2/12/105 indicates that the point connects to point 105 of station 12 on traverse 2.

TABLE 14.1.

First two digits refer to type of supplementary measurement

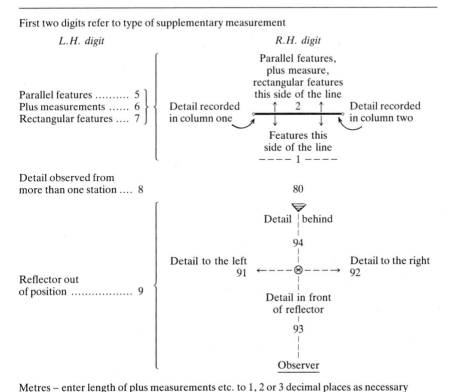

Metres – enter length of plus measurements etc. to 1, 2 or 3 decimal places as necessary

The central block is used for corrections to distance, where the prism is not actually held at the detail point, together with the supplementary measurements. The left-hand column is for the two characters representing the type of supplementary measurement while the right-hand column is used for the distance. For example, ⊗ R 3.69 indicates that the point of detail is 3.69 m to the right of the prism, or alternatively ⊗ + 1.50 indicates the point is 1.50 m beyond the prism. □ L 5.17, rectangular feature to the left, of 5.17 m side length. If the information is to be programmed to produce a set of rectangular co-ordinates for plotting, then an all-numeric system may be preferable, as shown in *Table 14.1*.

The fourth block allows first for a column for a point symbol to be used by the plotter to represent the detail or height point, then a column for a line type symbol, if any, and then a column for a relevant area symbol such as a roof, vegetation, slope, etc., followed by R or L if required to indicate right or left. Finally, the annotation column may be made use of to give additional information to clarify the data or point.

The reference may be very simple, such as just using the codes 1 to 9 for points and lines, with no area symbols or annotations at all. In this case, the area details and annotations would be added after the initial plotting

TABLE 14.2.

Point symbols		Line symbols		Annotations	
●	1	Solid line	1		
○	2	Pecked line	2	Boundary post	06
Single trees–coniferous	4	Elec. transmission line	3	Boundary stone	07
–others	5	Pipe line, at or close to GL	4	Cattlegrid	10
Culvert	7	–overhead	5	Chimney	11
Spot height only	8	Bank	6	Finger post	19
Bench mark	9	Drain	7	Flagstaff	23
		Direction of flow ←	8	Footbridge	24
		Line of constant slope	9	Footpath	25
Area symbols				Fountain	26
Archways 01, Pylons 02,				Gun	33
Roofs, glazed 03, others 04,				Issues	37
Sloping masonary-top 05, bottom 06,				Letter box	42
Steps ⨅⨅⨅ 07, ▤▤▤ 08,				Mast	48
Bog, marsh 10, Bracken 11, Bushes, shrubs 12,				Milepost	49
Furze, gorse, 14, Heath, heather, bilberry 15,				Milestone	50
Rough grassland 19, Undergrowth, underwood 22,				Monument	51
Orchard 16				Platform	60
Man-made slopes-top 80, bottom 81				Post ⎫ to prevent	61
				Posts ⎭ vehicular access	62
				Pump	63
				Spring	77
				Statue	78
				Tank	80
				Telephone call box	81
				Water tap	93
				Water trough	94
				Weighbridge	95
				Well	96

```
                    ┌─────────────────────┐
                    │  Present recorded    │
                    │  point of detail     │
                    └──────────┬──────────┘
                               │
   Area symbol    │    Area symbol
   to the left    │    to the right
   enter '1'      │    enter '2'
                    ┌──────────┴──────────┐
                    │ A previously recorded│
                    │   point of detail    │
                    └─────────────────────┘
```

stage. Alternatively, a more comprehensive coding may be used as set out in *Table 14.2*. This allows for 9 different point symbols, 9 line symbols, up to 99 area symbols and also an indication whether these are to be portrayed left or right of the line symbol. Similarly up to 99 annotations may be used. The number of possible codings could, of course, be greatly increased by adding alpha codings or using alphanumeric codes. The sample set of codings in *Table 14.2* are taken from a recent task and it can be useful to have the set in use printed on the reverse of the field sheet.

The final right-hand block of the form in *Figure 14.2* is used in heighting, to record the height of the reflector above or below the spot height being observed.

14.7   Office procedure

The computation of traverses, preliminary plotting considerations, the construction of grids, co-ordinate plotting, contour interpolation and plan completion after plotting are dealt with elsewhere in the text and will not be repeated here. This section deals with aspects directly relevant to bearing and distance survey office work, including the computations which may be required before plotting and the methods of plotting and drawing detail.

14.7.1   Calculations prior to plotting

Some or all of the calculations set out here may require to be carried out before or during the plotting of the survey.

14.7.1.1   Horizontal distance

Horizontal distance = slope distance × cosine *VA*,
±   a correction if the reflector is held in front of or behind the point.

14.7.1.2   Height difference

$\triangle h$ = slope distance × sine *VA*

14.7.1.3   Reduced level at reflector point

RL at point = station height $\pm \triangle h$ + instrument height − reflector height

If the plotting is to be carried out by rectangular co-ordinates and not by polars, then the following may be required.

14.7.1.4   Orientation correction for conversion of horizontal angular readings to bearings

Orientation correction = bearing (instrument to RO, from co-ordinates) − horizontal angular reading to the RO

14.7.1.5   Bearing to detail or spot point

Bearing to detail point = horizontal angular reading to detail point + orientation correction

14.7.1.6   Partial co-ordinates of line to a point

$\triangle E$ and $\triangle N$ obtained using P→R key, or
$\triangle E$ = horizontal distance × sine bearing and
$\triangle N$ = horizontal distance × cosine bearing

14.7.1.7   Point co-ordinates

E of point = E of station + $\triangle E$, and
N of point = N of station + $\triangle N$

14.7.1.8   Reflector offset correction

If the reflector is held to the left or right of the point (supplementary code 91 or 92 in *Table 14.1*) then the point co-ordinates may be calculated from a two-legged traverse. Bearing of the supplementary correction is bearing from station to reflector ±90°. If detail is to the right, code 92, add 90, if to the left, code 91, subtract 90.

Alternatively the station to reflector and reflector to point distances may be entered with the R→P key to deduce the station to point distance and the bearing correction.

14.7.1.9   Co-ordinate corrections for plus measurements, rectangular and parallel features

These may be carried out in a similar manner to the above.

14.7.1.10   Detail fix by intersection

Use the sine formula to calculate the unmeasured side lengths, then calculate co-ordinates by bearing and distance.

An alternative is to use the intersection formula, reference *Figure 14.3*.

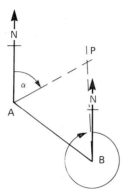

Figure 14.3

$$N_P = \frac{N_A \tan \alpha - N_B \tan \beta - E_A + E_B}{\tan \alpha - \tan \beta}$$

and

$$E_P = E_A + \tan \alpha (N_P - N_A)$$

or

$$E_P = E_B + \tan \beta (N_P - N_B)$$

Preferable to use bearings closest to 0°–180°

14.7.2   Plotting detail and spot height points

14.7.2.1   Plotting by polars (protractor or co-ordinatograph)

The protractor should be placed centrally over the station point, then oriented by reference to the observed readings to the RO and forward station. The protractor may be kept in place by drafting tape or paper weights. The polar co-ordinatograph will remain in place by its own weight.

Using the co-ordinatograph, each observed horizontal angular reading and its distance are plotted with the instrument's pricker. The point is circled in pencil, then annotated with its point number or, preferably, with its reduced level if the plan is to be contoured.

Using the protractor, plot all the bearings first, indicating each by a light tick in pencil and annotating their appropriate reference numbers. Remove the protractor and, with the aid of a scale rule, plot the required distance along each of the bearings. The plotted points should be circled and annotated as for the co-ordinatograph. Connecting all the bearings from the station mark to the plotted ticks by a pencil line should be avoided, since the station mark will be obliterated.

As mentioned previously, when plotting by polars the radius of the protractor must be at least as great as the length of the longest line to be plotted.

14.7.2.2   Plotting by rectangular co-ordinatograph

Due to the variation in models, the manufacturer's instructions should be referred to for guidance.

14.7.3   Drawing detail and contours

Using the scale rule and set square, plot all those points where the detail was to the left or right of the reflector, supplementary codes 91 and 92. Connect together all the points of detail as indicated in columns one and two of the team leader's form. Plot the remaining supplementary points of detail, codes 51 to 80. Pen in the detail in accordance with the symbols block, *Figure 14.2*. Interpolate the contours along all lines of constant slope, line symbol, code 9, *Table 14.2* and *Figure 14.2*. Interpolate the contours between all adjacent spot heights except between spot heights either side of a line of constant slope. Note that it may be more convenient to interpolate the contours on a separate trace of the area. See Chapter 9 for further information.

14.7.4   Plan completion

Plan completion is carried out as detailed in Chapter 4.

14.8   Microcomputer applications

Reference should be made to Section 12.4 for the MICROSURVEY programs relevant to the topics covered in this chapter.

Chapter 15

Photogrammetry

15.1 Introduction

Photography as applied to surveying is a vast subject and one aspect, *photogrammetry*, has become a very specialized surveying method, This is so specialized and expensive that very few firms practice the technique fully, despite its unique advantages. *Photogrammetry* may be defined as the science of obtaining reliable measurements from photographs, to determine the dimensions of photographed objects. Any individual involved in surveying should have an understanding of the technique, sufficient that he may know when it would be advantageous to make use of it and can call in specialists.

The photographic camera was first produced in 1839 and within a few years attempts were made to apply it to surveying. The first positive application was in the survey of high mountain areas, using what may be termed *Photographic plane tabling*. The important advances, however, have been in the application of *stereoscopic photography*, that is, two pictures of the same area from different viewpoints, principally in vertical photographs taken from an aircraft. This method permits the visual reconstruction of a three-dimensional image of the area viewed and thus maps may be made and also heights measured from photographs.

15.1.1 Classification of photographs

15.1.1.1 Air photographs

These are photographs taken from a camera in an aircraft in level flight. *Vertical photographs* are taken with the axis of the camera and lens vertical (or as near vertical as practical considerations will permit). These are the most important types of photographs from the surveying point of view. *Oblique photographs* are taken with the camera axis inclined to the vertical – *high oblique* if the picture includes the horizon, otherwise termed *low oblique*. Air survey cameras may be of plate or film type, according to specific requirements and in either case prints may be made on glass, paper or transparent film. Camera lenses may have an angle or field of view of

60° (normal), 90° (wide angle) or 120° (super wide angle) and the focal length of lenses may vary from 88 mm to about 210 mm. Picture sizes vary from about 140 mm square to 230 mm square. The typical application in planning and engineering work is a normal field lens, 152 mm focal length and 230 mm² paper prints.

15.1.1.2   Terrestrial photographs

These are photographs taken from the ground, with the camera axis horizontal or nearly so. They may also be termed *ground* or *horizontal photographs*. A *phototheodolite* may be used, an instrument combining a theodolite and a camera, one mounted above the other on a tripod.

Alternatively, a *stereometric camera* may be used to take two simultaneous pictures of an area from two positions. The Wild C120 is an example of this – two cameras with parallel axes, set on a bar and spaced 1.2 m apart, the whole mounted on a stand. This system permits a three-dimensional effect to be obtained.

15.1.1.3   Single vertical photograph

A single vertical photograph gives a map-like representation of the ground and it is of more or less constant scale over the whole photo area. If an area of land were covered with such photographs a map could be drawn from them, but it would be very variable in its accuracy with difficulties in making detail tally at the edges of adjoining photographs.

The scale of a single photograph may be determined approximately if the height of the aircraft and the mean height of the terrain above datum are known, together with the focal length of the camera lens. In *Figure 15.1(a)*, f = camera focal length, H = aircraft height above datum, h = mean height of terrain above datum, ab = length of photo, and AB = length of the area of terrain covered by the photograph. It will be evident that the scale ratio of the photograph is equal to ab/AB. However, since the triangles are similar,

$ab/AB = f/(H - h)$.

Assuming, as an example, flying height 550 m, ground height 100 m, f = 150 mm, then

photo scale = $0.150/(550 - 100) = 0.150/450 = 1/3000$, or 1:3000.

Distances on the photograph may be measured with a millimetre rule then multiplied by the scale factor to give ground distances. In fact, this only holds good if the camera axis is vertical, the terrain is level, all heights are correct and there are no distortions in the lens or film. In practice, an aircraft rarely flies straight and level, aircraft height is measured approximately by pressure variation, an altimeter, and the terrain is never completely level. Flying height cannot be guaranteed within about 10 m generally. *Figure 15.1(b)* shows the effect of tilt on a photograph and the changes of ground height in an area cause similar local distortions in the photo images. The view presented by a vertical photo is unfamiliar and it requires much experience to interpret correctly, but vertical photos form

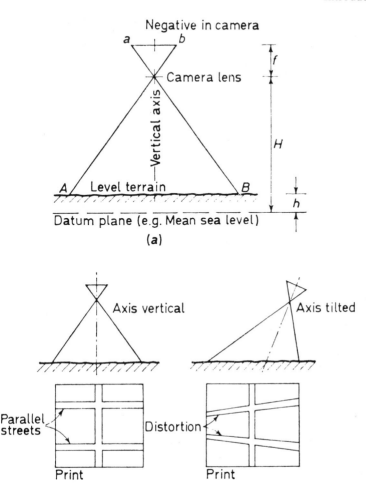

Figure 15.1

the best basis for map-making due to the comparative uniformity of scale as compared with an oblique and the use of stereoscopy for heights and contour information.

15.1.1.4 Stereoscopic pairs

Two photographs, from different positions, permit *stereoscopic viewing* of an area, i.e. viewing in depth or three-dimensionally. Special viewing equipment is normally required.

In terrestrial photography, the stereometric camera may be used, as mentioned earlier. In air photography, one vertical camera is used, in an aircraft in straight and level flight and successive photographs are taken at suitable intervals of time. The result is a series of overlapping pictures of the terrain and if an area of ground appears in the overlap of two photographs it may be studied stereoscopically. A pair of overlapping pictures like this is known as a *stereo pair* and they are generally simply two successive photographs out of a strip. Typically a 60% overlap is aimed at.

15.1.1.5   Oblique photographs

These give a more familiar view, almost a side elevation sometimes rather than a vertical photograph, but the scale varies over the photograph and distances cannot be scaled from prints. They are not as good as verticals for plotting purposes.

15.1.1.6   Colour photography

Colour is not used in photographic survey in the common sense, but see below.

15.1.1.7   Infra red photography

Infra red sensitive emulsion is used on occasion to give so-called false colours which aid interpretation of the photograph. For example, different types of trees and watercourses in woodland areas become easier to identify.

15.1.2   Survey applications of photography

The principal applications of survey photography are identified in the following sections.

15.1.2.1   Air survey and terrestrial photogrammetry

These are designed to replace much of the ground survey measurements by measurements from photographs, for the production of survey drawings in the orthographic and axonometric projections. *Orthographic projection* includes the normal maps, plans, sections and elevations. *Axonometric projection* is a three-dimensional effect in which both the plan dimensions and heights of buildings and other objects are true to scale. Such drawings are used then in the same way as traditionally produced drawings.

15.1.2.2   Information gathering from photo interpretation

A skilled photo interpreter may use photos to identify objects, their extent, significant characteristics, etc., including such things as likely crop yields, plant diseases, location of pre-historic remains, volumes of stock piles (e.g. coal), etc.

15.1.2.3   Monitoring of movement and other changes

The study of photos taken at intervals of time can allow the analysis and measurement of movement and other changes occuring over a period of time such as changes in beach formations, etc.

15.1.2.4   Preparation of photo-mosaics

Photographs covering an area of ground may be assembled together in their correct relative positions to form a *mosaic*, a kind of map. See p. 290.

15.2   Stereoscopy

15.2.1   Binocular vision and depth perception

Stereo is from the Greek, meaning solid, hence stereoscopy or seeing in three dimensions. In normal vision, when both eyes are focused on an object, the eyes turn inwards until their axes or sight rays intersect at the object sighted. This turning inwards is termed *convergence*, the angle between the sight rays is their *angle of parallax* or the *parallactic angle* and the distance between the eyes is their *interocular distance* or the *eyebase*.

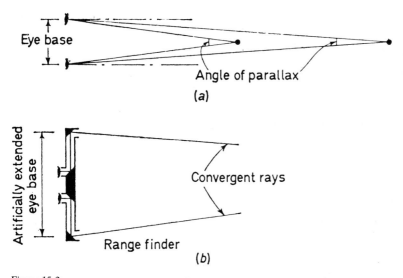

Figure 15.2

Figure 15.2(a) shows how the relative distances of objects from the viewer are judged by the magnitude of the parallax angles – the less the angle, the greater the distance and vice versa. The judging of the angle is done by the brain and when the parallactic angle is small (somewhere between 10 and 30″ of arc) the brain is unable to detect a change, therefore it is no longer possible to estimate distances or depth in this way. Note that this estimation of distance must not be confused with the estimation of distance by the relative apparent size of objects of known dimensions, that is to say, through experience. As an example of this latter, in the army, riflemen are told that at a certain distance a kneeling man will appear to be the same height as the rifle foresight blade. The eyebase may be increased artificially, thus increasing the parallactic angle of a distant object, as in binoculars or rangefinders (*Figure 15.2(b)*) or the mirror stereoscope.

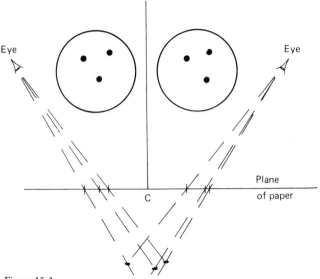

Figure 15.3

Figure 15.3 shows two circles, each containing three dots and these may be used to demonstrate stereoscopic viewing and depth perception as follows.

Place the figure flat on a table, hold a postcard vertically with its edge touching the page along the line marked PC. Position the head with the eyes vertically above the page, eyebase parallel to the line joining the two circles, eyes at normal reading distance from the page (wear reading spectacles if necessary). With the left eye vertically above the left-hand circle and the right eye vertically above the right-hand circle, slowly lower the head until the two circles merge – a slight rotation of the page may be needed to get this to happen.

Within a short period of time, between 5 and 20 s, the separately viewed dots should fuse together into a single set of three dots. Initially this may take some time to happen, since the eyes are not used to this form of viewing and with a very few individuals the images will not fuse.

When fusing takes place and the eyes have clearly focused on the dots, they will stand out sharp and clear. The right-hand dot should appear to be higher than the left-hand and both of these much higher than the third dot.

Stereoscopic viewing without a stereoscope is not easy and in this example the individual may initially see four, five or even all six dots, then suddenly they will all fuse together. When this happens, it will slowly become evident that the dots are at different depths. The lower part of the figure shows what appears to happen, with the lines of sight from both eyes converging on the dots at different depths.

15.2.2 Stereoscopic photo pairs

Two photographs from different positions give two views of an area in the same way as the two eyes get different views of an object they focus upon.

The two photographs may be viewed like the two sets of dots – one with each eye – and the result is a three-dimensional image of the area viewed. It is normally necessary to use special equipment to ensure that each eye sees one picture only and such devices are termed stereoscopes.

In vertical air photography, with two pictures overlapped by 60%, the overlap area of the photographs may be seen 'in depth', under a stereoscope. The *pocket stereoscope* is a small, cheap and handy instrument for the study of a stereo pair, the principle being shown in *Figure 15.4*. A and B are points of ground detail which appear on photo 1 as a and b and on photo 2 as a′ and b′. The lenses make the eyes act independently, thus the sight rays from the eyes pass through a and a′ to

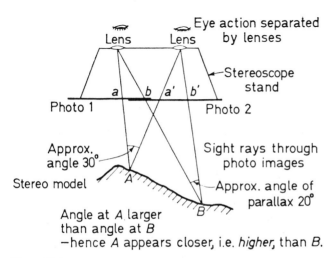

Figure 15.4

focus apparently at A. Similarly, when B is focused, the sight rays pass through b and b′. When the observer scans the overlap area in this way, he sees a three-dimensional image of the area. This 'solid image' is called a *stereo model* of the area, or simply a *model*. The pocket stereoscope is a most useful instrument for rapid study of a pair of photographs in depth – the impression of depth assists in the identification of photo features which are difficult to recognize in the single vertical photograph.

15.2.2.1 *Orienting photos under the stereoscope*

To orient the photos under the stereoscope for viewing, proceed as follows:

(1) Place the photographs in sequence as they were taken.
(2) Separate the photographs, along the flight line, so that common detail features are approximately 60 mm apart.
(3) Place the stereoscope on top of the photographs in such a way that each lens is over the same point of detail on the respective prints.

(4) Holding the stereoscope and the left-hand print firm, look through the stereoscope lenses and move the right-hand print about until the images fuse into one. When this occurs, the area viewed should appear as a three-dimensional model.

Any area of the overlap may now be studied by moving the stereoscope over it with parallel shifts. It may, however, be necessary to adjust the right-hand print by a small amount. Note that to scan the whole of the overlap it may sometimes be necessary to fold up the edge of the upper photograph, which may obscure the view of part of the lower photograph. If shadows are present on the photographs, viewing will normally be better if the photos are oriented so that the shadows appear to fall towards the observer.

15.2.2.2 Magnification

To see all the detail on a good quality photography clearly it should be observed through a lens system which gives at least $5\times$ magnification. This is due to the fact that the naked eye does not have the visual acuity to separate very close detail on a photograph.

15.3   Aerial photography and mapping

15.3.1   Advantages and disadvantages

The method is clearly advantageous if site access is restricted. It is cheaper for the survey of large areas at small to medium scale mapping and is often faster than ground survey methods. Setting up costs are very high and some air survey companies find it cheaper to use EDM methods for 1:500 scale mapping. Some tasks, particularly small areas, are cheaper by alternative methods and then air survey companies will prefer to use these other methods.

15.3.2   Mapping methods available

Vertical photography may be used to provide maps, plans and other survey data by a variety of methods. These include:

Photo mosaics.
Tracings from air photos.
Plotting from stereo pairs using stereographic plotting machines.
Determining data by analytical methods using the mathematical relationships between the object and its image, including production of rectangular co-ordinates, angles, distances, heights, volumes, from scaled measurements or analytical stereo plotters.

15.3.3   Flying for photographic cover

When an area is photographed for mapping, the camera aircraft flies in a straight line at a pre-determined height, taking successive overlapping photographs at timed intervals. The result is a strip of photographs covering a strip of terrain. If one strip is not enough to cover the terrain required, the aircraft turns and flies parallel strips as necessary. In order to avoid gaps in the photography, the fight paths are arranged so that adjoining strips overlap one another by 10, 20 or often 30%. This is termed *lateral overlap, Figure 15.5* showing the result of these overlaps in diagrammatic form. Note that if the photography is good, some areas of the terrain may appear on six photographs.

Camera type, lens, flying height, air base (distance B between successive camera positions) and aircraft speed are specified according to the purpose of the photography, photo scale required, etc. Very small-scale mapping is often done with super wide-angle lenses now, these giving very large ground cover and economy of flight expenditure. Large-scale work, on the other hand, is of limited area and needs large-scale prints and normal lenses are generally used.

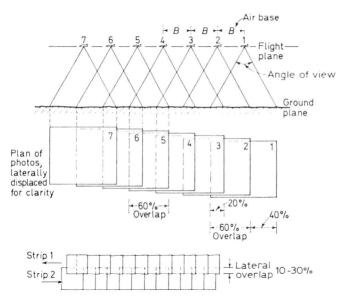

Figure 15.5

When a photograph is taken, the camera should also record on the negative the altitude above MSL, the date and time, the focal length of the lens and the serial number of the particular photograph, together with four or more *fiducial* or *collimation marks*. These collimation marks act, for the photograph, like the cross-hairs in a telescope. They may be located in the middle of the sides or in the corners of the photograph and if the imaginary lines connecting opposite marks are drawn, then their intersection indicates the *principal point* or geometrical centre of the print.

15.3.4 Ground control, completion

An air survey, relying on photographs which may contain errors due to tilt and relief, must be controlled by a network of ground control points which have been accurately surveyed by normal ground survey methods. These points must, of course, be of such a nature that they are clearly visible on the photo prints.

The ground survey methods used depend upon the type of area and the accuracy specified, typically traversing or triangulation for plan control and spirit levelling or trigonometrical heighting or other methods for height control.

Normally, it is necessary to provide a minimum of three height control points per photo overlap area and two plan control points, but four fully co-ordinated points are preferable. This amount is reduced if the technique of *aerial triangulation* is used during the plotting process. Ground control points may be natural or artificial features, according to circumstances and if surveyed before the photography it is termed *pre-pointing*, if after the photography it is *post-pointing*. Artificial marks must, of course, be pre-pointed.

An individual photo negative may be corrected (*rectified*) so that it fits the known three-dimensional relationship of the ground control points, thus scale errors due to tilt are eliminated and those of relief reduced. Apart from ground control, the ground surveyor must generally visit the survey area after photography, in order to check on detail hidden by overhanging trees, clouds lower than the aircraft, etc., and to investigate detail the plotter cannot identify. Local names and similar information must also be collected.

15.3.5   Mosaics

When air photos covering an area are assembled together in their correct relative positions, the result is termed a *mosaic*. If the photographs have not been rectified, there will be discrepancies in the detail at the photo edges, this being regarded as an *uncontrolled mosaic*. If the photographs have been rectified to fit the ground control it is a *controlled mosaic*. Sometimes a mosaic is controlled by mounting the prints on a map of the area and adjusting the print detail to fit the map detail.

Mosaics are useful in many applications, but they do not give stereo effect and are more difficult to read than a map – they contain much detail which is ignored when a map is compiled, and they do not show place names or symbol information. Single air photos or mosaics can be reproduced by lithography when many copies are required. There is some loss of contrast but it does enable names and symbols to be added, to produce what is sometimes known as the photomap.

Orthophotography and *orthophotomaps* are recent developments. The former is photography which rectifies the original perspective projection to an orthographic projection by removing the effects of camera tilt and ground height variations. An orthophotomap is produced from one or more *orthophotos*. This form of rectification is known as *differential rectification* and instruments for this are often linked to stereographic plotters, as below.

15.3.6   Plotting from air photographs

The production of accurate maps from the stereoscopic study and analysis of air photographs is the province of the photogrammetric specialist. Aerial survey and mapping requires a large team of specialists and the use of very large, complex and expensive plotting machines. In the typical modern plotting machine, the *stereographic plotter*, the stereo pair is viewed stereoscopically and the operator looks through binocular eyepieces to view a black dot which appears to move over the stereo model. Various controls allow the dot to be moved along the outline of photographic detail, the movement being reproduced to an appropriate scale by a plotting arm which moves a pencil over a plotting table.

The dot may also be made to appear to move vertically and its height can be read off a scale in the machine. If the dot is set at the height of a required contour, it can be set to just appear to touch the surface of the ground in the model, then if it is kept at the contour height and moved along the ground surface it will trace out the required contour. The contour, of course, will be drawn out by the pencil on the plotting table.

It is possible, however, for the non-specialist himself to prepare a plan of a small area, such as a construction site, using simply a stereo pair and some comparatively inexpensive equipment. The simplest method is probably using the *radial line plotter* to plot planimetric detail, together with the determination of relative heights from the photos using a *parallax bar* and a stereoscope. These methods will not be described here, their use being unlikely to be economical except in unusual circumstances.

The early stereographic plotters were described as *analogical* stereoplotters, later instruments often being known as *analytical* stereoplotters. The latter instrument is, in fact, a combination of a *stereocomparator*, a computer and an automatic drawing table. The stereocomparator is designed to measure the full co-ordinates of any point on the overlap of a stereo pair, the automatic drawing table is often a rectangular co-ordinatograph controlled by the computer. The computer is complete with keyboard, disk drive and printer.

15.4   Terrestrial photogrammetry

Terrestrial photogrammetry is a method of survey using ground photography in which the end product, the drawings or rectified photographs, are used by building surveyors, architects, archaeologists, etc. It is a satisfactory alternative to the traditional survey methods for those engaged in the recording, analysis, conservation and restoration of buildings, historical sites, monuments and other three-dimensional objects. The technique is used by engineers to monitor movement, determine earthwork quantities, etc., and by others not involved in land surveying. For example, in the motor industry, medicine, agriculture, road traffic accident investigation, fashion design and so on. It is primarily a task for the photogrammetric specialist, and although the fieldwork, the photography and the survey control may be carried out by a surveyor, the plotting must be done by the specialist.

15.4.1   Advantages and disadvantages

15.4.1.1   Advantages

As in air survey, direct access to the object being surveyed is not necessary, hence problems impossible by traditional methods may be tackled, e.g. the survey of unstable structures. Photogrammetric plotting from stereo pairs allows the most complex detail to be drawn without difficulty. All the information may be collected at the one instant in time, relevant data being abstracted as needed, even after the object has been destroyed. The minimum of time can be spent on site, sufficient only to establish control and take the necessary photographs.

15.4.1.2   Disadvantages

As with air survey, the principal disadvantage is the initial cost of the equipment. A suitable camera will cost more than a theodolite and the plotting instrument is still more expensive.

15.4.2   Available methods

This form of survey may be based on the use of one or more individual photographs, or, as in air survey, overlapping stereo pairs.

15.4.2.1   Individual photographs

The accuracy of single photo work is not as high as can be obtained with stereo pairs, but satisfactory results are possible if a good quality camera and enlarger are used. The best results are obtained with a *metric camera*, designed for use in terrestrial photogrammetry and with a lens capable of minimum distortion on the negative and a high resolving power.

The ideal use of a single photograph is where the face of the object being photographed is flat, as in many building facades. The camera should be set up with the plane of the film parallel to the face of the object. Scale may be obtained from counting bricks, by taping a distance on the object then scaling it on the print, or by placing a levelling staff or ranging rod against the face of the object prior to photography.

This method imposes severe limitations, including the problem of camera tilt, likely distortion at the edge of the film with some cameras, the skill needed by the enlarger operator to achieve the correct scale and the depth of the object with respect to scale and focusing.

Rectification of a single photograph of the flat face of a building is possible. This may be done graphically by the use of a perspective grid on the photograph, optically with some form of projection, photographically with a photographic rectifier, or analytically using the mathematical relationships between the object and the image. Mosaics produced from single rectified photographs printed on to transparent film and from which dyeline prints can be made is a method in use. The accuracy is not of the best, but it is often adequate for the client's needs.

15.4.2.2   Overlapping stereo pairs

As with air photography, the process with stereo pairs is the production of drawings using analogical and analytical stereoplotters and by orthophotography. Control is normally by co-ordinating points on the object or building by intersection or by EDM, and sometimes as in Section 15.4.2.1 above.

Problems which may be minor in vertical photography may be significant in terrestrial work. These include:

The inability of the camera to maintain accuracy, being unable to focus over the possible range of object distances.

The inability of early stereoplotters to cope with the range of object distances and the camera tilt. Plotters designed specifically for terrestrial work have now been developed.

On site, possible short distances between the camera and object may arise, due to obstructions such as narrow streets, making it necessary to tilt the camera or take many photographs. For this reason platforms are often erected for camera stations, or hydraulic hoists are used.

The difficulty of providing control on tall, thin structures.

An important limitation is the ratio of the photo scale to the survey scale. The ideal ratio is 1:5, a possible maximum 1:10. Thus if the survey scale is to be 1:50 then the photo scale must be not less than 1:500. The photo scale is the ratio of the focal length to the object distance, e.g. a 60 mm focal length and a 30 m object distance gives a photo scale of 1:500 (60:(30 × 1000)), see *Figure 15.1(a)*. Taking photographs across a 6 m wide road, the photo scale might be 60 mm/6 m or 1:100.

A second limitation is the ratio of the base line and the object distance. A recommended ratio is 1:10, thus for an object distance of 30 m the approximate distance between camera stations should be 30 m/10 or 3 m.

With cameras designed for terrestrial photogrammetry a typical subject area covered might be 40 m × 50 m at an object distance of 30 m and with a 3 m base line an overlap of 37 m × 50 m. However, in a narrow street where the object distance is 6 m, with camera stations 2 m apart (a possible minimum with separate stations), much of the foreground will be lost and the overlap will be reduced to roughly 6.6 m × 6.0 m.

An alternative approach when the stereo pair are oblique to one another is to use what has been called *graphical stereorestitution*, using scaled measurements on the photographs with respect to their principal points. The method has been out of favour for many years but has recently been revived under the name of photo-radiation for the monitoring of beach movements in Wales. A stereocomparator or similar instrument is used to measure the photographic distances (see the Bibliography).

15.5   Airphoto interpretation and maps

The study of air photographs for the identification of ground features and the acquisition of information is termed *photographic interpretation*. This is a highly skilled art, requiring much experience and practice, since the view presented by a vertical photograph is quite unfamiliar to the eye. It is a little easier if stereo pairs are used under a pocket stereoscope so that the ground relief shows up. Photographs are usually studied for a specific purpose, such as urban development, construction works, etc., and it is essential that the interpreter be knowledgeable in the particular field of study concerned.

In attempting to identify an object on the photographs, the interpreter should consider (1) the *shape* or *pattern* of the photo image, (2) its *size*, (3) what shape and size of *shadow* the object casts, (4) the *tone* and *texture* of the photo image, and (5) whether the object appears to be one of two or more *commonly associated objects* or *features*.

The distinctive *plan shape* of some objects may be sufficient to identify them, e.g. churches, road junction roundabouts. *Elevation shape* is sometimes visible and may be decisive, e.g. the waisted shape of a cooling tower as compared with the shape of an oil storage tank. Regularity of *pattern* would distinguish an orchard from a small wood. If the photo scale has been calculated, it will be possible to scale off the actual *size* of objects viewed, thus size would differentiate between a circular dug well and a circular sewage filter bed. *Shadows* may confirm or disprove the deductions from shape and size, as with the shadow of a dome which otherwise might be mistaken for a gasholder or storage tank. Tree types, deciduous or coniferous, will be clear from shadow forms.

The *tone* of an image indicates the amount of light reflected back to the camera, thus a smooth surfaced asphalt road will appear light in tone as compared with a newly-chipped road surface. If a particular tone is associated with an identified object it may help to indicate more of the same features. *Texture*, or variation in tone, is also important. The colour of an object has less effect on the tone of its image than the nature of its surface.

The *common association* of certain forms or structures may assist in identification, particularly if the nature of one of them has already been established. Thus a series of circular and rectangular shapes outside a community is often the local sewage works, graveyards normally lie beside churches, etc.

When studying photo interpretation, it is instructive to compare a stereo pair and a map of the same area. Most people are familiar with maps and find them easy to read, because

(1) a map is an abstract, showing only that detail which is thought necessary for its particular purpose,
(2) place names, street names and in large-scale maps even individual house numbers are shown,
(3) the use or nature of many buildings and features is shown by symbols, or words, e.g. churches, windmills, post offices, etc., and
(4) factual height information is shown by spot heights above datum, contour lines, hatching and shading, etc.

On the other hand, a map may be difficult to read because it does not tally with the known ground – maps are always, in fact, out of date, since years may elapse between ground survey and actual map production.

By contrast, air photographs show a wealth of confusing detail not normally shown on maps, such as individual bushes and trees, crops, haystacks, varied ground tones due to varying moisture content, vehicles on roads, etc. In addition, there are no names or symbols to assist identification of buildings and ground features and there is no height information. Air photographs are up to date at the instant of photography, and prints can be available for study within a few days, thus changing conditions can be studied almost as they occur.

Table 15.1 shows some of the important differences in detail on a stereo pair and on a map.

TABLE 15.1.

Detail	On stereo pair	On map
Water	May be black, white, or any tone between. Depends on depth, silt, weeds. Clear water dark, muddy water light	Symbols, words or coloured
Streams and rivers	Irregular outline. Ground relief indicates fall	As above. Direction of flow shown by arrow
Canals	Uniform width, regular shape, runs along a contour line	Named. Locks shown
Woodlands	Trees dark in tone. Shadow indicates type. Trees may be counted if photography carried out during winter. Trees and bushes differ in size. Heights of trees may be judged	Symbols for types. Often no indication of density. No heights
Uncultivated land	Varying tone according to nature and relief	Symbols for some areas – oziers, etc.
Cultivated land	Short grass/crops light in tone. Taller crops darker. Different crops shown by tone variation. Ploughed fields have regular dark tone	Not shown
Orchards	Regular pattern differentiates from woodlands	Symbols
Ground relief	Clearly visible in stereoscope	Contours, spot heights. Symbols for cuttings, embankments, tunnels, etc.
Land drains	May show by relief and effect of ground moisture content on tone	Not shown
Footpaths	Show clearly in grass. May not show in sand or under tree cover	Symbol sometimes
Roads	General light tone. Footways, verges, ditches, culverts, bridges, all visible	Symbols may show class and route number
Railways	Long straights and smooth curves as compared with roads. Rails may be counted	Symbols
Buildings	Building size and relative heights may be estimated. Nature may be judged from surroundings such as paths, gardens, industrial areas, etc. Elevation may be visible	Plan dimensions may be scaled. No heights or elevation information. Use may be indicated in words or symbols

TABLE 15.1.(*cont'd*)

Detail	On stereo pair	On map
Sewers	MH covers and street gullies visible on large-scale photographs	Not shown
Boundaries	Hedges and fences visible may judge type by elevation or shadow. Legal boundaries not shown	Single line only, no indication of type. Administrative boundaries may be shown by symbols
Overhead wires and cables	May be visible in elevation, shadows of posts-pylons useful. Height may be judged	May or may not be shown by symbols
Temporary conditions	Changes in land use, areas under floods, storm damage, size of stacks of stored materials, size of spoil tips, variation in excavation, progress of construction works, traffic flow on roads	Not shown
Seasonal conditions	Crop distribution, moisture content variation, forestry development, etc.	Not shown

Chapter 16

Large-scale map revision

16.1 Introduction

Large-scale map revision is a graphical method for the supply of detail, the detail typically being plotted in the field on a copy of an existing plan of the survey area. It is used where the detail on an existing map is to be updated, or where a map or plan has been produced but the survey detail is incomplete. Generally the task is carried out by an individual rather than a team, using a minimum of equipment and with the minimum of detail recorded. The method is accordingly cheap and can be accurate if the plan being revised or completed is accurate.

Essentially the three basic techniques for supplying detail are used (bearing and distance, intersection, chaining), adapted to a graphic approach.

16.1.1 Application

In addition to the revision of the detail on existing maps, this method may be used to 'field complete' plans produced by air or ground survey methods. Large-scale air survey plans are rarely fully completed from the air photographs, since detail may be obscured by cloud cover, foliage, tall buildings, roof overhangs, etc., and names and similar information must be collected on site.

In a built-up area, particularly where the detail is close, it is unnecessary and unwise to fully complete a chain or EDM survey during the execution of the survey. Completion by an individual using map revision techniques will be more economical and allow the main survey party to proceed to other tasks. Further, the completion by map revision methods provides a form of independent check.

16.1.2 Equipment

The plan which is to be revised, completed or checked must be attached to a suitable size drawing board, preferably with a protective cover. A3 is an appropriate size. The following items of equipment will be required:

A straightedge, attached to the board in a sleeve.
150 mm 60° set square.
150 mm scale rule.
Suitable grades of pencil, with protective caps.
Pencil eraser.
Pencil sharpener, pen knife or razor blade.
Pocket size notebook.
Double prism optical square (referred to simply as an 'optical square' here).
20 m synthetic tape, complete with hook and skewer at the zero end.

The 20 m tape is preferred, since it winds into a smaller case than the 30 m tape and may fit the hip pocket. The shorter tape also discourages the user from taking undesirably long measurements.

The tape hook and the skewer serve as the surveyor's assistant, providing a fixing for the tape end when the surveyor is making single-handed measurements. A 100 mm maximum length meat skewer is suitable, looking rather like a small arrow and suitable for sticking in the ground, jamming in masonry joints, etc.

The equipment is light, small and easily carried with minimum effort. The board may be kept under the arm, the pencils, set square and scale in a breast pocket, the tape in one hip pocket and the notebook, eraser and pencil sharpener in the other. The optical square may be kept on a string around the neck when in use.

16.1.3   Care of equipment

The synthetic tape should be checked periodically against a steel tape, possibly twice a week if in daily use. The scale should be checked for straightness and the right-angle of the square checked. Straightedge, scale and square must be kept clean, with an occasional wash in soapy water. Pencils must be kept constantly sharp and protective caps used. Ball point pen caps may be used.

16.2   Requirements of the plan or map for revision or completion

The majority of the existing detail portrayed on the map or plan should be to an acceptable standard of accuracy. Thus accuracy across the plan should be correct to at least one part in 500 and scaling between two adjacent points of detail should be correct to 0.15 m at 1:500 scale, 0.75 m at 1:2500 and in similar proportion at other scales. Inaccurate control encourages a surveyor to lose confidence in this excellent method of supplying detail.

The plan or map should be covered with a network of control, *control* in this case being any accurate point on the plan which can be identified on the ground. Before the commencement of revision the surveyor will be uncertain as to which detail is correct, but he should assume that all detail is correct until it is shown to be otherwise.

The density of the control will depend upon the amount of detail portrayed, but ideally there should be a network of points not greater than about 30 m apart at 1:500 scale and roughly proportional at other scales, e.g. 150 m at 1:2500 scale. If the density is less than recommended, then it may be increased by graphical intersections, as shown below.

16.3   Field methods

16.3.1   Terminology

Known detail is that which is at present shown on the plan. *New detail* is that which has yet to be supplied.

New detail is surveyed from the control or known points, which may be traverse, triangulation or chain survey stations, or chain survey tie points, or known detail. In other words, new detail may be surveyed from any points which can be identified both on the ground and on the plan.

Detail is supplied from a combination of lines, distances or both, mainly by intersection or by bearing and distance. A line is either a shot, a straight or a production.

A *shot* is a line of sight between two known points, generally passing through one or more other points of detail. A shot is sometimes known as a ray. If two points of detail are identified on the ground and on the plan, the surveyor may place himself at a point on the shot between the points using an optical square and sighting the two points.

A *production* is the prolongation of a line of sight.

A *straight* is the extended alignment of any straight detail.

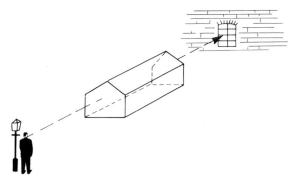

Figure 16.1

Figure 16.1 illustrates a straight feature (the long face of the building) and its extended alignments. If the surveyor is standing at the lamp-post and has just located the straight (i.e. he is standing on the straight) then it is said that the surveyor has *picked up the straight*. If the surveyor sights along the straight and it appears to intersect the centre of the window, then the surveyor is *shooting the straight* and the straight *shoots on to* the centre of the window.

A *distance* is a taped measurement taken between two known points. If a line is measured throughout its length, known point to known point, then it has been *tied out*. If the line is only partially measured and the overall length is therefore unchecked, it is not tied out and is described as being an *indirection line*.

The intersection of two or more lines is known as an *intersection*. Accordingly, a point where three lines cross is sometimes known as a *trisection*.

16.3.2   Methods of supplying detail

16.3.2.1   *Intersection of two or more lines*

Points of detail may be fixed by the intersection or two or more lines, the lines being shots (connecting two detail points) or productions (the extension or prolongation of a line). As mentioned above, the surveyor may use the optical square to position himself on a shot, or line himself in on a production. Where two rays are used to fix an intersection then it should be checked by a taped measurement to existing detail, or by a third line of sight. An intersection point having been marked on the ground, the same lines may be drawn on the plan to fix the intersection point on the drawing. The intersected point, fixed both on the ground and on the plan, may serve as a new control for the fixing of further detail.

16.3.2.2   *Short measurements along indirection lines or productions*

The application of these is best explained by an example. Consider a new asymmetric roundabout, constructed at the junction of five roads and it is required to survey the new roundabout to update the plan of the area. A possible procedure is as follows.

Fix a point roughly central in the roundabout, by intersecting two rays between existing detail. Mark the point on the ground and plot it on the plan, checking it by taping to another detail point. Anchor the zero end of the tape at the intersection using the skewer. Walk to the kerb, unwinding the tape. Identify a point of existing detail, then standing at the kerb move left or right until the detail point and the intersection are in line and measure the production of that line from the skewer to the kerb. Draw the line on the plan and its production, then scale off the taped distance along the production and plot the kerb point.

Repeat the process around the kerb of the roundabout, selecting suitable detail points, until the whole of the roundabout has been plotted. Measurements from the intersection point to the kerb, along a line towards another detail point (an indirection line) may be used similarly.

As a general rule, taped measurements for indirection lines should not exceed one tape length and for productions they should not exceed the lesser of one tape length or one-third of the length of the production. Fixing by production tends to be weak and important firm detail supplied by production should be independently checked by other means.

16.3.2.3   *Longer running measurements between control*

Detail may be fixed by taking running measurements along a line joining two control points. Such lines should always be tied out and generally they should not exceed 30 m in length at 1:500 scale, or 60 m at 1:1000 and pro rata at other scales.

Longer taped lines like these are often used to pick up straights (extensions of straight lines of detail, such as the face of a building) and fix the ends of the straights. It is preferable to avoid using offsetting, since the technique is slow and really needs two surveyors and two tapes.

New buildings may often be surveyed for plan updating by taking the external straights of the building (its external wall faces) and extending these until they meet the boundary detail. The points where the straights shoot the boundaries may be surveyed and plotted, then running measurements taken along the straights to fix the position of the building. When the major shapes have been fixed, the minor detail on site may be picked up by short measurements along productions or indirection lines.

16.4   General

Map revision is essentially visual and graphic, taped measurements being used only for short distances between known points and along known alignments. As with all surveys, no two surveyors are likely to carry out a revision task in exactly the same way, but the methods should accord with those described in the previous sections. The chosen approach should aim to produce an accurate result in the least time and with minimum expenditure of effort. It should not be necessary to walk over the same ground several times and ideally the procedure should be such as to involve walking over any area of ground once only.

Independent checks are essential, but must not be overdone – it should be remembered that control is assumed correct until proved wrong. The work should be checked as it proceeds, using a combination of known and new detail to provide further new detail.

Excessive taping and booking are to be avoided, booking generally only being required when measuring lines with many readings. The surveyor should be able to memorize four or five readings before plotting them. Booking must be kept simple and preferably the booked detail plotted within an hour of measuring. Single line booking as for the skeleton lines in chain or building survey is adequate. Short measurements along productions and indirection lines should be plotted immediately after measurement. Where a straight shoots on to a known building, it is quicker if the surveyor can avoid the need to actually go to the building to measure the distance to its corner or other known point. The brick counting technique may be used, if appropriate, or the building may be visually divided into proportional parts by reference to the windows, etc.

Where a row of buildings such as a block of lock-up garages or terraced houses appear to be all the same size, the block should be measured overall and then divided proportionally into individual units. The individual properties should not be measured, but it is advisable to measure the end units, since these tend to differ in size from the remainder.

At all times the survey principles of working from the whole to the part, the independent check, consistency and the safeguarding of the field document must be remembered. One object of a survey is to produce a plan which the surveyor, the client and the users believe to be correct. Thus, adjacent elements of detail must be seen to be in agreement with one another as regards shape, size, proximity and direction. Buildings which are symmetrical and rectangular on the ground must appear so on the plan, since deviations from symmetry on a plan are immediately evident to the eye. Similarly, straight lines on the ground must be straight on the plan.

Where a measured line has been tied out and it does not scale correctly within an acceptable tolerance, or if two adjacent straights in a rectangular building are not at right-angles on the plan, or a trisection is not a perfect intersection, then an error exists and must be located. If a check reveals no obvious errors in scaling or plotting, then it is likely that there is an error in one of the known points which has been used.

A shot taken across and roughly at right-angles to a measured straight line may assist in providing a check measure on the line, hence determining whether one end of the line is in error. Where two adjacent straights on a building are in doubt, a third straight could be taken along another face, or a third line of sight could be taken to cut the original straights near their intersection, to provide a check. A doubtful intersection or trisection may be checked by running another line from independent detail, or by taking tape measurements from other detail.

Finally it may be noted that paper shrinks and stretches, therefore it is advisable to check the true scale of the plan or map before commencing revision. Scale changes must be allowed for in the plotting and scaling of distances.

16.5   Office procedure

New detail surveyed and any corrections or deletions to existing detail should be inked-in daily, since otherwise the new information may be lost among the construction lines drawn on the plan. Corrections and deletions may be indicated by small crosses, but this depends upon the method in which the plan is to be produced for the client. A revision task may involve completely re-drawing all detail, or it may only necessitate making amendments to earlier reprographic material. Again, the original may be stored in a digital format.

Part 5

Setting out

Chapter 17

Setting out

17.1 Introduction

Setting out is the name used for all the operations required for the correct positioning of works on or adjacent to the ground, together with their three-dimensional control during construction. Setting out operations are the reverse of normal survey, since the dimensions are known but the points have to be located on the ground or on the structure.

Setting out for traditional building work does not require high accuracy, but with the development of precisely dimensioned frames and components it is frequently found that insufficient attention has been paid to the initial site dimensioning and the consequent poor fit of beams, panels, etc., may be extremely expensive.

Although the setting out of every job is different, certain jobs are standard and standard solutions are available for many tasks. The various requirements in setting out are considered here under four main headings – plan control, height control, vertical alignment control and excavation control. The headings are self-explanatory, but excavation control combines all the others in a specialized aspect.

17.1.1 Setting out documents and data

The documents and data available to the setting out surveyor or engineer may vary from a simple verbal communication to a most complex array of plans and specifications. Setting out plans or drawings may include *block plans, site plans, location drawings* and *detail drawings* and between them they should identify the site, relate new work to any existing and include the fundamental setting out data, i.e. the positions to be occupied by the various spaces and elements of the buildings or works and their dimensions.

In addition there may be a specification, shedules, survey data, possibly computer printout and mathematical tables. A *specification* describes material and workmanship, but it may contain information relevant to the setting out, such as the depth of bed for drain lines. A *schedule* tabulates information on numerous and repetitive items and again there may be relevant data such as the heights and invert levels of manholes. The *survey data* relates to the original site survey and may include information on traverse stations, bench marks, etc. *Computer printouts* may be included with the survey data and are often used where the position of detail for setting out is defined by two- or three-dimensional rectangular co-ordinates. Computer printouts are tending to replace the use of specialist mathematical tables for the setting out of high-speed rail and motor roads. See the Bibliography for more advanced texts on these matters.

17.1.2   The essentials of efficient setting out

These may be considered to be *accuracy, timeliness* and *clarity*. Lack of *accuracy* can be costly in time and money, since time must be wasted in rectifying mistakes. Where the surveyor makes frequent mistakes, there will be a loss of confidence by the construction team who will then waste additional time in unnecessary checking. *Timeliness* involves not merely setting out to programme, but having the setting out completed early and in advance.

If there is time in hand, then it will be possible to cope with emergencies such as vandalized marks, unforeseen changes in the construction programme, etc. If the setting out is early, then the surveyor need not work in haste or under undue pressure and both of these tend to lead to undetected mistakes.

Clarity means that all the personnel involved should be able to understand the surveyor's setting out and the meaning of all pegs and other marks defining the location and height of the new works. Colour coding of pegs and marks will assist in this aim. Again, although setting out is the reverse of surveying, it requires a skilled application of survey techniques and adherence to the principles of survey covered in Chapter 1.

17.1.3   Stages in setting out

17.1.3.1   Document inspection

The relevant documents must be examined and checked for any information relating to the setting out, including any existing site survey data and all must be checked for adequacy and accuracy.

17.1.3.2   Site inspection

The nature of the site must be examined and the existence of obstacles to measurement and ranging noted, bearing in mind the methods and equipment to be used. The nature of the ground will affect the type of markers to be used and also the reliability of existing survey marks. Existing control must be verified, including bench marks and any changes noted.

17.1.3.3   Proving the site drawings

The size and shape of existing detail on the site must be checked against the drawings and it must be verified that the new spaces and elements will fit on to the site in the manner shown on the drawings. As an example, the sum of individual measurements along a line must be compared with the overall length of the line, gradients must be checked against the actual levels shown, etc. Map revision methods may be used to check existing detail. Temporary bench marks should be checked by re-levelling from the O.S. bench marks.

17.1.3.4   Preparing setting out data

Any calculations such as bearings and distances, co-ordinate transformations, curve computations, etc., should, as far as possible, be carried out before commencing the task on site.

17.1.3.5   Site setting out procedure

Assuming the use of traditional methods, with distances set out by tape and angles by theodolite, the detailed *plan* of the new works is set out first. When this is complete, suitable *check measurements* must be made, according to the circumstances. For example, a rectangular plan may be checked by calculating the lengths of diagonals then measuring them and comparing the actual and theoretical distances. When the plan is correct, *heights* may be fixed by ordinary levelling from the verified temporary bench marks.

If EDM equipment is in use, it is possible to set out each point in all three dimensions, using the angular bearing, horizontal distance and vertical height difference. This must be checked, but gross errors in line and level may be checked by eye, short check measurements may be taped and heights can be checked by levelling. Another form of check is to set out the same point twice, from two different instrument positions.

17.1.3.6   Setting out profiles, etc.

If detail points are set out on the faces of new buildings, or on the centre-lines of roads, etc., such marks will be destroyed when the construction work commences. Accordingly, it is customary to establish a *reference frame* of accurately located lines outside the actual construction and clear of all obstructions, then the new works are located by measurement from the reference frame as required. Typically, in a small building, this is arranged by setting up *profile boards* opposite the ends of each wall, then marking the positions of the extended wall lines on these boards, as in *Figure 17.1*.

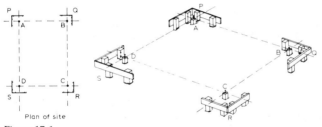

Plan of site

Figure 17.1

On large buildings and similar structures, the reference frame may consist of two lines at right-angles to one another, set out by theodolite. A better form, however, is a frame of four lines located by theodolite, outside and parallel to the proposed structure. Pegs may be placed on these lines to mark, for example, the *centre-lines* or *faces* of rows or lines of stanchions, columns, or walls to be erected.

Offset pegs are used to indicate the lines and heights of kerbs, drains, sewers and so on. On occasion offset points may be set out directly by the surveyor or engineer, but sometimes they may be set out by site operatives, from the data supplied by the surveyor.

17.2   Plan control

Plan control covers the setting out of detail in two dimensions in plan. Any pair of points may be related then either by a bearing and a distance or by a difference in Eastings and a difference in Northings (rectangular co-ordinate methods). Plan control may be supplied at any height, either at ground level, at the bottom of an excavation, or at intermediate floor levels, as required.

17.2.1   Methods available

New detail may be located from

existing detail,
existing control,
new control tied to the existing control,
a grid superimposed over the site, or
a combination of any of the above.

Where a *grid* is used, it may be the grid of the existing control and survey, i.e. a *survey grid*, or it may be a special site grid. A *site grid* may exist on the site drawings, that is pre-contract planned and this may or may not be the survey grid. Alternatively the site grid may be positioned by the setting out surveyor or engineer on the drawings to coincide with the orientation of the major plan units shown.

Whatever the new site detail is set out from, the setting out will be the basic survey methods of supplying detail, which are rectangular offsets, bearing and distance, intersection, or a combination of these. The method to be used should be that which will be the most economical in time and money, provided that the required accuracy can be achieved.

In the United Kingdom the traditional approach is by *rectangular offsets* since this is most commonly the form in which the information is presented. On small sites the offsets are typically from a framework of lines tied by map revision techniques either to survey control (if any) or to existing detail.

On larger sites, the framework may be a *grid superimposed over the site*. In its simplest form this can be just the four corners of a rectangle, but it may be a series of grid points around the perimeter of the site from which other points and ordinates may be located. Detail is located either by intersection of ordinates or by short offsets from them.

Following the introduction of scientific calculators, setting out by *intersection of bearings* from two or more instrument stations became an economical alternative and, with the development of short range EDM equipment, setting out by bearing and distance also.

In some parts of the world property boundaries are defined by co-ordinated points and where the parcels are small, and many rectangular in plan, much of the setting out of the boundary points may be carried out by intersection of ordinates from a controlling traverse framework.

With modern sophisticated equipment it is feasible to set up an instrument close to the estimated position of a required point, then provided several traverse or similar co-ordinated stations are visible the instrument station co-ordinates may be deduced by resection or other control link methods. The co-ordinates of the required point being known, it may then be set out from the co-ordinated instrument station by bearing and distance.

The methods to be used will depend upon the equipment and labour available, the skills of the operators, the survey data provided, the size and nature of the job, the topography of the area, the method of presentation of the information and, of course, the accuracy of the site plan.

17.2.2 Equipment

The equipment used for setting out may include all the measuring and note-taking equipment mentioned earlier in the text. Ground marking materials are needed, particularly timber pegs, stakes and battens, nails, steel reinforcing rods, cement and sand, chalk, crayons, paint and tools such as hammers, hand saws, spades, cold chisels, centre punches, etc. A builder's line (propylene twine), plumb-bobs and profiles or profile boards (see *Figure 17.1*) may also be needed.

17.2.3 Common setting out tasks

From the surveying point of view, the actual operations involved in plan setting out reduce to

setting out a horizontal angle
lining-in a point between two existing marks
prolonging a straight line
setting out an exact horizontal distance, and
setting out a curved line in plan.

17.2.3.1 Setting out a horizontal angle

To set out an angle using a glass arc theodolite, the method is essentially an application of the *simple reversal* technique used to measure horizontal angles. The procedure is as follows

(i) Set the theodolite over the point on the line at which the angle is to be raised, centring and levelling up as usual.

(ii) On *face left*, sight on to a target placed on the far end of the line (the *reference object*) and read and book the horizontal circle reading. Any specified zero may be set on the circle before sighting the RO, but generally a zero of 00° 00′ 00″ is used.

(iii) Release the upper or horizontal plate and telescope clamps and turn the instrument through the exact desired angle, directing an assistant to place an arrow or other mark at the required distance on the alignment defined by the telescope cross-hairs.

(iv) Repeat the whole operation on *face right* using a new zero in a different quadrant of the horizontal circle.

(v) If the two marks do not coincide, the mean position between the two indicates the required alignment, provided that both have been placed with equal care. A large difference indicates gros error, of course.

Building Research Establishment Digest 234, 1980, *Accuracy in Setting-Out*, quotes a probable accuracy of ±5 mm in a distance of 50 m using a 20″ glass arc theodolite centred by optical plummet, and ±2 mm in 80 m using a similar 1″ theodolite.

Right-angles may be set out using a steel tape to construct a 3:4:5 triangle and a large wooden right-angled triangle frame is often used on small building jobs. The optical square may be used for rough preliminary work, but it is not suitable for setting out an accurate right-angle.

17.2.3.2   Lining in between two existing marks

If chain survey accuracy (±50 mm) is adequate for the task, as it may be on rough preliminary location work, then the methods described under chaining in Section 4.5.1 may be used. For normal accuracy setting out, a theodolite is used for lining in. Depending upon the inter-visibility of the end points and also on whether it is possible to set the theodolite up over an end point, three distinct cases may arise.

Case 1 is where the end points are intervisible (visible one from another) and the theodolite can be set up over one end.

Case 2 is where the end points are not intervisible but both are visible from some intermediate point, or alternatively where the end points are intervisible but the theodolite cannot be set up over an end point.

Case 3 arises where there is no intermediate point at which both of the end points are visible.

Case 1 solution: Set up the theodolite over one end point, sight on the other end point, then direct an assistant to place a mark at the required intermediate position.

Case 2 solution: The procedure is as follows

(i) Use the chain survey techniques of 'lifting a line' (Section 4.5.1.2) to locate the approximate position of an intermediate point on the line, such that both ends of the line are visible from the intermediate point.

(ii) Set up the theodolite over the approximate point, levelling and centring as usual.

(iii) On *face left*, sight on to the more distant of the end points and bisect the target carefully, then *transit* the telescope and direct an assistant to place an arrow exactly on the line of collimation and in the vicinity of the nearer end point.

(iv) Keeping the instrument on *face right*, turn the telescope, sight on the more distant point again, bisect the target carefully, then transit the telescope and place another arrow on the collimation line and close to the nearer end point.

(v) The mean position of the two arrows now defines the alignment from the far end point through the instrument station. Measure the offset from this alignment to the near end point.

(vi) Calculate the proportional distance the instrument must be moved in order to bring it on to the correct alignment joining the end points.

(vii) Move the instrument the required distance, repeat the observations, and repeat as necessary until the instrument is on the correct alignment. In the early stages it may be necessary to move the tripod, but later it will be possible to simply slide the theodolite over the tripod head into its correct position.

Case 3 solution: If the end points are not intervisible and there is no intermediate point from which both ends can be viewed, as for example in a wooded area, then a possible solution may be to run a traverse between the end points but as close to the alignment as possible. If the traverse is co-ordinated, then the co-ordinates of the required intermediate point may be computed and the point set out by bearing and distance from the nearest traverse station.

17.2.3.3   Prolonging a straight line

This task may, again, be carried out by chain survey methods if that accuracy is sufficient, but in site setting out it is normally necessary to use a theodolite. Assuming a line AB is to be extended to the point C, two cases arise.

Case 1 arises where the theodolite can be set up on A and the position of C is visible. All that is necessary is to set up the instrument on A, point towards B, bisect the target carefully, finishing with the horizontal clamps applied, then re-focus and raise or lower the telescope and direct an assistant to place a mark on the collimation line at C.

Case 2 is where the theodolite cannot be set over A, or the position of C will not be visible from A. In this case, set up the instrument over B, sight the target at A on face left, transit the telescope and direct the assistant to place a fine mark exactly on the collimation line in the vicinity of C. Repeat the process on face right, then in the absence of gross error the mean of the two marks is on the required line.

17.2.3.4   Setting out an exact horizontal distance

In setting out an exact horizontal distance, the alignment is fixed by theodolite and the required distance is then fixed by steel taping or EDM. When using a steel tape, two cases arise, *Case 1* where the distance is less than one tape length and *Case 2* where the distance exceeds one tape length.

Case 1 solution: Set the theodolite over the start point, place a peg on the required alignment using the methods described above, then measure the desired distance along the line to the peg and move the peg as necessary. The peg alignment position and the distance must be checked alternately, as needed, until the peg is at the correct distance and on the true alignment. In building work it is common practice to use large section wooden pegs, with a small nail placed to protrude vertically from the top of the peg. The nail provides a fine mark for sighting and also for measuring to and a suitable mark for centring a theodolite over.

Case 2 solution: Set the theodolite over the start point, place a temporary peg accurately on the desired alignment and within 3 or 4 m of the required distance (approximate taping only). Now measure the distance to the temporary peg carefully and calculate the additional distance required. Move the theodolite and set it up over the temporary peg, then set out the additional distance to the final peg in the same way as shown for Case 1 above. Finally carry out a check measurement of the distance from the start point to the final peg to ensure that the peg is at the required position.

Where distances are to be set out by simple EDM equipment, a similar approach may be used. Thus the reflector is placed on line at the approximate distance, the distance to the reflector determined accurately and the added distance to be set out is calculated and fixed by a short taped measurement from the reflector position. A peg may be located at this new position, then the reflector placed at the peg and the distance determined as a check.

17.2.3.5   *Setting out a curved line in plan*

The most common form of horizontal curve in practice is an *arc of a circle*, termed a *circular curve*. A *simple circular curve* is an arc of constant radius, joining two tangents. A *compound circular curve* consists of two simple curves, of different radii, meeting on a common tangent and having their centres lying on the same side of the common tangent. A *reverse circular curve* is similar to a compound curve, but the two simple curves have their centres on opposite sides of the common tangent.

Other curve forms may be encountered occasionally, such as the *ellipse* or the *parabola* and these, together with short radius simple curves, may be marked on the ground using chalk, crayon, scratch marks, etc., the curves being defined by templates or by locating centres or focii and swinging arcs of a length of tape or cord. The simplest example is a simple circular arc which may be fixed by locating the centre of the circle, placing a peg there, then swinging the desired radius from the centre. Similar techniques may be used for ellipses, etc., and arcs may also be marked by driving pegs at intervals along their length.

Roads, railways and canals have plan alignments consisting generally of straight lines connected by simple circular curves of very large radius and the methods mentioned above are impractical. In these cases, the positions of points on the curve must be calculated and then set out in the field using steel tapes (linear methods) or the theodolite and steel tape or EDM equipment. The curve is set out, in effect, by a series of straight chords, the lengths of the chords being typically not greater than one twentieth of the radius of the curve, so that there is very little difference between the chord and arc between any two adjacent points. Setting out by these methods is termed *curve ranging*, and it is treated separately in Section 17.3.

On roads and railways where high-speed traffic isinvolved, a *transition curve*, that is to say a curve of constantly changing radius, may be introduced between the straight and the simple curve so as to allow a gradual change of direction and to reduce the effects of centrifugal force by a gradual application of super-elevation. Transition, reverse and compound curves are not covered in this text.

17.2.4   Use of rectangular co-ordinates

Rectangular co-ordinates, introduced in Chapter 6, are very widely used in plan setting out works today. All the methods of Section 6.6 are relevant, including co-ords from bearing and distance (P→R), bearing and distance from co-ords (R→P), co-ordinate grid transformation and all the control link methods such as resection, etc. The use of these methods will be assumed where relevant and they will not be treated further here.

17.3   Curve ranging

Curve ranging will be considered here on the assumption that it is aimed at setting out roads, but the same principles are, of course, applied in railways and the setting out of large radius curves generally. Although the traditional method for planning roads is to lay out straight lines in plan then connect them together with *circular* or *transition/circular/transition curves*, it should be noted that the more modern method is to treat the road alignment as a continuous curve (a cubic spline) and compute the rectangular co-ordinates of points at appropriate intervals long the road line. The mathematics of this method are somewhat advanced and formerly required the use of a large computer. Although this is now accessible to microcomputer users, it will not be covered here.

17.3.1   Terminology

Figure 17.2 shows two straight lines, ab and bc, representing the centre-lines of two successive road straights intersecting at b, and a circular arc connecting the lines. The circular arc is part of the circumference of a circle, the lines ab and bc are tangential to the circle and they are tangents which touch or meet the circle at a and c.

The following list defines the various parts of the figure, together with the name and symbol used for each in practice. Note that there is some lack of agreement in terminology between English-speaking practitioners.

Distance ab = bc = tangent length = T
Distance da = dc = de = radius of the curve = R
Arc length aec = length of the curve = L
Distance eb = external distance = E
Distance ac = long chord
Distance ef = mid ordinate = M

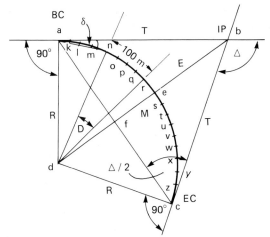

Figure 17.2

Angle $\triangle$ = deflection angle of the curve
        = angle adc at centre of circle
Angle D = degree of the curve,
        = angle subtended at the centre by an arc of 100 m
Angle abc = supplement of $\triangle$,
        = intersection angle
Angle bac = angle bca = $\triangle/2$ = total deflection angle
Angle ban = tangent deflection angle of the arc an, from point a
        = angle at a between the tangent and the chord an = δ
Angle bad = angle bcd = 90°
(angle between tangent and radius drawn to the point of contact)

Assuming a right-hand curve, moving from a towards c,
Point a = tangent point 1 = TP1 ⎫
        = beginning of curve = BC ⎬ Alternatives
        = point of curvature = PC ⎭
Point c = tangent point 2 = TP2 ⎫
        = end of curve = EC ⎬ Alternatives
        = point of tangency = PT ⎭

Point b = point of intersection = IP
Point d = centre of the circle
Point e = midpoint or crown point of the curve

    Road centre-lines are usually marked with pegs at uniform intervals and when the pegs are identified by their distance from the start of the job it is known as *through chainage*. The *uniform interval* is a sub-multiple of 100 m, commonly 20 m. The convention for expressing chainage of a peg is to state the number of hundreds of metres followed by a plus sign and then the remaining distance, e.g. a peg at 17 120 m from the start has a chainage of 171 + 20.0. The *tangent points* are unlikely to be at exact multiples of 20 and could have values such as 171 + 22.5 and 173 + 13.1, i.e. 17 122.5 m and 17 313.1 m from the start. The lower value is of course at the beginning of curve BC, while the greater is at the end of curve EC.

Figure 17.2 shows how the curve may be marked by pegs at these *standard intervals*, forming a set of *standard length chords*. The first and last chords will actually be *sub-chords* of less than the standard length.

17.3.2 General circular curve calculations

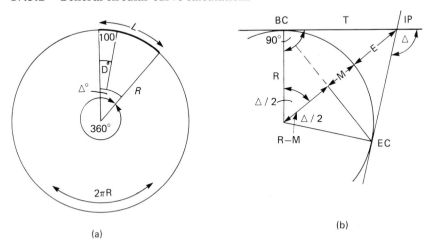

(a)

(b)

Figure 17.3

17.3.2.1 To calculate curve radius, given degree of curve (Figure 17.3(a))

$$2\pi R/100 = 360°/D°,$$

therefore

$$R = 36\,000/2\pi D$$
$$= 18\,000/\pi D \tag{17.1}$$

17.3.2.2 To calculate degree of curve from given radius

$$D = 18\,000/\pi R \tag{17.2}$$

17.3.2.3 To calculate tangent length (Figure 17.3(b))

$$T = R \tan (\triangle/2) \tag{17.3}$$

17.3.2.4 To calculate the curve length (Figure 17.3(a))

$$L/2\pi R = \triangle°/360°$$

therefore

$$L = 2\pi R\triangle/360$$
$$= \pi R\triangle/180 \tag{17.4}$$

or

$L/100 = \triangle°/D°$

therefore

$L = 100\triangle/D$ (17.5)

17.3.2.5   To calculate external distance (Figure 17.3(b))

$\cos(\triangle/2) = R/(R + E)$

therefore

$R + E = R/\cos(\triangle/2)$

and

$E = (R/\cos(\triangle/2)) - R$

therefore

$E = R((1/\cos(\triangle/2)) - 1)$ (17.6)

17.3.2.6   To calculate the mid ordinate

$\cos(\triangle/2) = (R - M)/R$

so

$R \cos(\triangle/2) = R - M$

and

$M = R - R \cos(\triangle/2)$
$ = R(1 - \cos(\triangle/2))$ (17.7)

17.3.2.7   To calculate the long chord length (Figure 17.3(b))

$\sin(\triangle/2) = $ half length of long chord$/R$

so

long chord $= 2R \sin(\triangle/2)$ (17.8)

17.3.2.8   To calculate tangential deflection angles for arcs

In *Figure 17.4*, a chord of length C is shown, starting at a tangent point (the beginning of curve in this case). If the *tangential deflection angle* (the angle between the tangent and the chord) of the ars is δ, then the angle subtended by the arc/chord at the centre of the circle is 2δ. (A geometrical theorem states that the angle between a tangent and a chord to the tangent point is equal to half the angle subtended by the chord at the centre of the circle.)

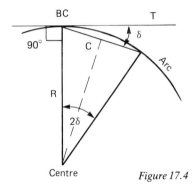

Figure 17.4

From *Figures 17.3* and *17.4*,

2δ/arc length $= \triangle/L = D/100 = 360/2\pi R$

so

$$\delta \text{ (degrees)} = \text{arc length} \times \triangle/2L \tag{17.9}$$
$$= \text{arc length} \times D/200 \tag{17.10}$$
$$= \text{arc length} \times 90/\pi R \tag{17.11}$$

Alternatively,

$$\delta \text{ (minutes)} = \text{arc length} \times 1718.87/R \text{ minutes} \tag{17.12}$$

This latter form is the most useful in setting out. Note that the tangential deflection angle varies directly with the *arc* length, thus if a *standard arc length* is used a *standard angle may be calculated* and for a multiple of the arc length the same multiple of the angle can be used. In practice, if a standard chord is $= <R/20$, then for all practical purposes the chord and its arc may be considered to be of the same length.

Referring again to *Figure 17.4*,

$$\sin \delta = C/2R \tag{17.13}$$

therefore

$$C = 2R \sin \delta \tag{17.14}$$

and this may be useful in various applications.

17.3.3   Linear equipment methods

Three main methods are available for setting out large radius curves using linear measuring equipment only – deflection distances, offsets from tangent and offsets from the long chord. In each case one or both of the tangent lines and tangent points must be located on the ground before the curve can be pegged and the radius must be known. Setting out commences from a known tangent point.

17.3.3.1 Deflection distances method (Figure 17.5)

A standard chord length of C must be calculated, $= <R/20$, of suitable length. In the figure, a standard chord runs from the tangent point a to the first point on the curve b, the angle between the chord and tangent is δ and the angle at the centre of the circle is 2δ. The perpendicular distance from the point b to the tangent line is termed the *offset distance*. Given R and C, the offset distance must be calculated.

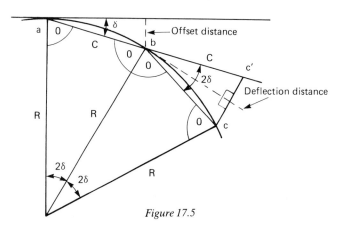

Figure 17.5

Now,

$\sin \delta =$ offset distance$/C$,

therefore

offset distance $= C \sin \delta$,

but from Equation (17.13),

$\sin \delta = C/2R$,

therefore

offset distance $= C^2/2R$ (17.15)

In practice, C and the offset distance are calculated, then the first curve point is fixed by swinging a tape or chain length of C from the point a, until the required offset distance from the tangent is achieved.

To set out the next curve point, the first chord must be extended (on the same alignment) by a further distance C and a peg inserted at c′, then the chain swung towards the curve by a distance of twice the offset distance, locating the new point c on the curve. The distance c′c is termed the *deflection distance*, calculated from

$\sin \delta =$ deflection distance$/2C$

therefore

deflection dist. $= 2C \sin \delta$

but

$$\sin \delta = C/2R$$

therefore

$$\begin{aligned}\textit{deflection dist.} &= 2C^2/2R, \\ &= C^2/R \\ &= 2 \times \text{offset distance}\end{aligned} \qquad (17.16)$$

All subsequent points on the curve may be located in the same way as point c.

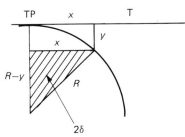

Figure 17.6

17.3.3.2 Offsets from tangent method (Figure 17.6)

The figure shows a tangent to a curve, the tangent point and the radius at the tangent point. $x = $ a distance measured along the tangent from the TP, and $y = $ the offset from the tangent to the curve. In the shaded triangle,

$$(R - y)^2 = R^2 - x^2,$$

therefore

$$R - y = \sqrt{R^2 - x^2}$$

and

$$y = R - \sqrt{R^2 - x^2} \qquad (17.17)$$

Thus, for any distance, x, along the tangent, an offset distance y, may be calculated.

Where it is required to set out pegs at specified distances along the curve, using the central angle 2δ of the triangle,

$$\sin 2\delta = x/R,$$

or

$$x = R \sin 2\delta \qquad (17.18)$$

and

$$\cos 2\delta = (R - y)/R$$

therefore

$$R \cos 2\delta = R - y$$

and

$$\begin{aligned}y &= R - R \cos 2\delta \\ &= R(1 - \cos 2\delta)\end{aligned} \qquad (17.19)$$

The value to use for 2δ is determined from the specified chord length, as shown earlier. Alternatively, it may be required to set out the curve by dividing its length into a number of equal chords, and 2δ may be deduced from the deflection angle of the curve and the number or chords.

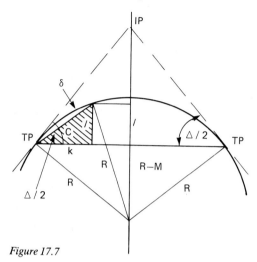

Figure 17.7

17.3.3.3   Offsets from long chord method (Figure 17.7)

In the figure, *l* is an ordinate from a point on the long chord to a point on the curve, while *k* is the distance along the long chord from a TP to the ordinate. In the shaded triangle,

$$\cos(\triangle/2 - \delta) = k/C \qquad \text{and} \qquad \sin(\triangle/2 - \delta) = l/C$$

but Equation (17.14)

$$C = 2R \sin \delta$$

therefore

$$k = 2R \sin \delta \cos(\triangle/2 - \delta) \tag{17.20}$$

and

$$l = 2R \sin \delta \sin(\triangle/2 - \delta) \tag{17.21}$$

Again, δ may be deduced as required as in the preceding method.

If arbitrary values of *k* are selected, then Pythagoras' Theorem may be used to calculate the appropriate *l* values.

$$(l + (R - M))^2 = R^2 - ((\text{long chord}/2) - k)^2$$

or

$$l = \sqrt{R^2 - ((\text{long chord}/2) - k)^2} - (R - M) \tag{17.22}$$

If the long chord and mid ordinate values have not been calculated, then from Equations (17.7) and (17.8)

$$l = \sqrt{R^2 - (R \sin(\triangle/2) - k)^2} - R \cos(\triangle/2) \tag{17.23}$$

17.3.4 Tangential deflection angles method

17.3.4.1 Principle of the method

This method, the most common for large radius curves, requires the use of a *theodolite* to set out *deflection angles* from the *tangents*, together with a chain or tape to set out *short standard chord lengths* along the curve. Generally the standard peg interval of 20 m is used for these chords and the even chainage points are carried through so that there are sub-chords at the beginning and end of the curve. As stated earlier, if the standard chord is not greater than one-twentieth of the radius, then the standard chord and its arc may, for practical purposes, be considered the same. If the standard chord must be less than 20 m, then it is preferable that it be a sub-multiple of 20 m.

The essence of the method is that successive arcs are measured out from a tangent point, while the deflection angle for each arc is set off from the tangent line. Strictly speaking an arc cannot be set out directly so instead the arcs are built up by setting out successive standard chords, each approximating to a short arc.

The curve should be set out in two halves – one half set out from the BC, the other half from the EC – and ideally they should meet at the *crown*

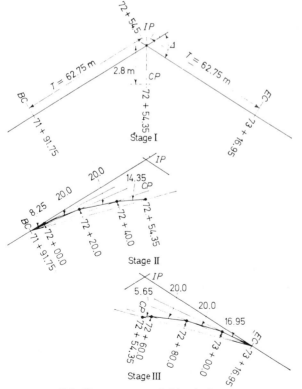

Note: Diagram exaggerated for clarity

Figure 17.8

point CP. The CP should, therefore, be located and pegged before setting out the two halves and the inevitable small errors will show up by the two halves failing to close on to the CP. A small discrepancy is to be expected and, provided it is reasonable, a few pegs in each half are adjusted, the adjustment diminishing with increase in distance of a peg from the CP. If the misclosure is large, gross errors should be looked for, such as a missed deflection angle. The acceptable misclosure cannot be defined, being dependent upon the specification, the equipment used and experience.

17.3.4.2   Example curve calculation

Curve No. 5R (meaning fifth curve along the road and curving to right-hand side)
Required radius = 699.80 m
Centre-line pegs at 20 m. Chainage of IP from the start, 72 + 54.5. $\triangle$ measured, found to be 10° 15′.

Calculation sheet (distances to 0.05 m)

IP	= 72 + 54.5	$\triangle$ = 10° 15′
R	= 699.80 m	$\triangle/2$ = 05° 07′ 30″
Chords	= 20 m	$\triangle/4$ = 02° 33′ 45″

$T = R \tan \triangle/2 = 62.75$ m
$L = \pi R \triangle/180 = 125.20$ m
$L/2 = 62.6$ m
E (external distance) $= R ((1/\cos (\triangle/2))-1 = 2.80$ m

Chainages

IP	= 72 + 54.50			
$-T$	0 + 62.75			
BC	= 71 + 91.75			
$+L/2$	0 + 62.60			
CP	= 72 + 54.35	Check	BC	= 71 + 91.75
$+L/2$	0 + 62.60		$+L$	1 + 25.20
EC	= 73 + 16.95		EC	= 73 + 16.95

Centre-line pegs required at	*chord lengths*
BC 71 + 91.75	0
72 + 00.0	8.25 sub-chord
72 + 20.0	20.0
72 + 40.0	20.0
CP 72 + 54.35	14.35 sub-chord
72 + 60.0	5.65 sub-chord
72 + 80.0	20.0
73 + 00.0	20.0
EC 73 + 16.95	16.95 sub-chord
	Total 125.20 $=L$, check. $\sqrt{}$

Deflection angles required, from $\delta = 1\,718.87\,l/R$ minutes

$$\text{for } l = 8.25, \quad \delta = \frac{1718.87}{700} \times 8.25$$

$$= 2.456 \times 8.25 = 20.26' = 20'\ 15.6''$$

for $l = 20.0$, $\delta = 2.456 \times 20.0 = 49.12' = 49'\ 7.2''$

for 14.35, $\delta = 2.456 \times 14.35 = 35.25' = 35'\ 15.0''$

for 5.65, $\delta = 2.456 \times 5.65 = 13.88' = 13'\ 52.8''$

for 16.95, $\delta = 2.456 \times 16.95 = 41.63' = 41'\ 37.8''$

Setting-out table for curve
First half, theodolite set over BC

Peg sighted	Chord	Circle reading
IP	0	00° 00′ 00″
		+ 20 15.6
72 + 00.0	8.25	00 20 15.6
		+ 49 07.2
72 + 20.0	20.0	01 09 22.8
		+ 49 07.2
72 + 40.0	20.0	01 58 30.0
		+ 35 15.0
CP→ 72 + 54.35	14.35	02 33 45.0 ≈ △/4 check √

Second half, theodolite set over EC

Peg sighted	Chord	Circle reading
IP	0	360 00 00
		− 41 37.8
73 + 00.0	16.95	359 18 22.2
		− 49 07.2
72 + 80.0	20.0	358 29 15.0
		− 49 07.2
72 + 60.0	20.0	357 40 07.8
		− 13 52.8
CP→ 72 + 54.35	5.65	357 26 15.0
	Check	360 00 00
		−357 26 15.0
		02 33 45 = △/4. √

17.3.4.3   Field operations

The field operations for setting out the typical curve above fall into three stages, as shown in *Figure 17.8.*

Stage 1: Fix the IP by projecting the tangent straights to intersect and then measure angle △ and find the chainage to the IP. The radius or degree of curve being specified, calculate all the items needed for setting out, as shown. Measure distance T from the IP and fix the BC and EC points on the tangents. Fix the CP by bisecting the angle BC–IP–EC and measuring out the distance E.

Stage 2: Set the theodolite over the BC point, orient it on the line to the IP (reading zero when sighted on the IP). As a check, set out the angles $\triangle$ /4 and $\triangle$/2 from the tangent in turn, they should bisect the CP and EC pegs respectively. Set on the first deflection angle of 00° 20′ 16″ and direct an assistant to place a peg on the sight line at distance 8.25 m from the BC. Mark the peg with the chainage, 72 + 00.0. (Note that although the angles have been shown to decimals of a second in the calculation, the rounded values to be used will depend upon the actual instrument in use.)

Set on the next total deflection angle 01° 09′ 23″, direct the assistant to pull the chain on and insert the next peg on the sight line at a distance of 20 m from peg 72 + 00.0, mark the new peg 72 + 20.0.

The procedure is repeated until the sub-chord to the CP is set out.

Stage 3: Move the theodolite to the EC point, orient it on the IP. Repeat the previous procedure, swinging the theodolite anti-clockwise and deducting the deflection angles from 360°, until the CP is reached again.

17.3.5   Two theodolites method

This method simply uses a theodolite placed over each tangent point, then peg locations are fixed by the intersection of their tangential deflection angles from both tangents. The calculation is carried out in the same way as in Section 17.3.4.2 above, but the Setting Out Table for the first half is extended to cover all points from BC round to the EC and the second half of the table is extended to cover all points from the EC round to the BC.

To fix a point on the curve, the two theodolite operators must collaborate carefully, taking it in turn to sight an assistant holding a ranging rod, until the rod is precisely on both sight lines for the two angles to the point.

17.3.6   Polar co-ordinates method

If a theodolite can be placed at the centre of the circle and all points on the curve are visible from the centre, the chainage pegs may be located on the curve by *polar radiation*. The curve essentials must be computed as before, but instead of deflection angles the *central angles* for all the chords must be calculated then successively added. (The central angle for a chord is equal to twice the tangential deflection angle for the chord.)

To set out the curve, the theodolite is set at the centre, oriented on the BC point, then the first central angle is set on and an assistant with a ranging rod directed on to line at a distance of R from the instrument. The next angle is set on, the assistant directed on to line again at distance R from the instrument, and so on. Since R is likely to be some hundreds of metres, this method is not suited to steel tapes but is ideal with EDM equipment designed for setting out, such that R can be measured out directly.

17.3.7   Rectangular co-ordinates method

This method is suitable where the whole of a job is being carried out on a rectangular co-ordinate grid. The road tangent lines may be defined by either three or four sets of co-ordinates and then it is possible to compute rectangular co-ordinates for all the points on the curve (IP, BC, EC, CP, etc.) and for all the standard chainage peg positions. Given the rectangular co-ordinates of all these points and the rectangular co-ordinates of selected instrument stations, polar co-ordinates of the points may be computed and set out from the instrument stations using suitable EDM equipment. The calculations will not be covered here, the reader is advised to re-read Section 6.6, some are rather complex and better handled by programmable calculator or microcomputer.

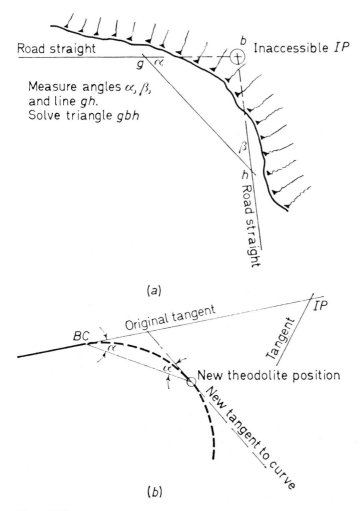

Figure 17.9

17.3.8   Common problems

A variety of problems may crop up in practice, but two common ones are worth mentioning. Where the IP of a curve is inaccessible, such as on tight curves on mountain roads, or where there are buildings on the line, a straight line may be laid between the two road straights and its length and the angles it makes with the straights measured. Thereafter, the inaccessible triangle at the apex may be solved and the distances from the IP to the cross-line calculated. After calculating the tangent length, the distance to BC and EC from the cross-line can be worked out and the points located as normal. See *Figure 17.9(a)*.

Another common trouble is that when setting out from one end it may not be possible to sight right through to the CP. The theodolite should then be moved to the last point sighted on the curve, set up there, then oriented so that when the reading on the circle is zero the telescope is sighted along the *tangent to the curve at that point*. Thereafter deflection angles may be set out again in the normal way. See *Figure 17.9(b)*.

17.4   Height control

The control of *heights* or *levels* in construction work is generally a simple matter. Accurate temporary bench marks should be fixed on site at points where they are unlikely to be disturbed, fenced off if necessary. The TBMs should be levelled from national or municipal bench marks, such as OBMs and a careful record kept of their reduced levels. Thereafter the height of any part of the site or the construction may be determined by reference to the site TBMs, using the surveyor's level and staff, or EDM equipment with trigonometric heighting, or a rotating laser giving a reference plane, or specialist site operative's equipment such as the water level or the site operator's automatic level.

When using the surveyor's level it may occasionally be necessary to fix heights above the level of the collimation line and in this case the staff may be used *inverted*, the readings then being *negative* values. This is detailed in Chapter 8.

Where the point to be heighted is more than a staff length above the collimation line, then a white steel tape may be pulled taut downwards from the point, the tape being read in the same way as a staff. An alternative method is *trigonometric heighting*, but this is not generally as accurate as levelling, though it may be adequate for some tasks.

17.4.1   Accuracy of levelling in construction works

If permissible deviations or accuracies are not specified for a task then ISO document 4463–1979 or BS 5964: 1980 may provide guidance. As an example, these indicate that when check levelling between a given TBM and an OBM they should agree within 10 mm, while in checking points on steel structures the height differences should agree within 2 mm.

17.4.2   Equipment

All the heighting equipment previously referred to may be used in site levelling or height control, including surveyor's levels, EDM equipment, lasers and all their ancillary equipment, together with pegs, profiles, etc.

The *site operative's automatic level*, not previously described, is small, fast, convenient for small jobs, easy to use and inexpensive. Due to its lack of magnification the sighting range is normally limited to 30 m and the manufacturers quote an accuracy of 6 mm at this range. The instrument has a pendulum mirror system, defining a horizontal collimation line when approximately levelled on a light-weight tripod. It is used with a special staff carrying a horizontal cross-bar capable of being moved up or down the staff. The eyepiece on top of the instrument presents two images of the target bar, separated by a vertical line. The operator signals the staff holder to move the cross-bar vertically until the two images are symmetrical and meet at the central vertical line. When this condition is attained, the cross-bar is at the same height as the optical centre of the instrument and the staff holder reads the vertical distance from the peg to the bar on a graduated scale on the back of the rod. Although cheap compared with a surveyor's level, the instrument is rarely seen on site now, due to the range and accuracy limitations.

Lasers, again not previously described, were introduced into the construction industry some twenty years ago and have become established as efficient setting out instruments for both height and alignment control. Laser is an acronym for Light Amplification by Stimulated Emission of Radiation and a laser projects a narrow beam of light, providing an alternative to the visible builder's line or the line of sight through an optical instrument. The line of the beam, depending upon the application, may be horizontal, vertical, or at any desired gradient. In some instruments, the beam of light may be rotated at speed, thus establishing a visible horizontal or vertical reference plane.

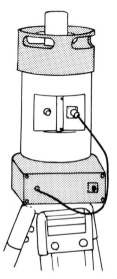

Figure 17.10

Figure 17.10 illustrates such an instrument, the AGL Beam Machine, height approximately 340 mm, with a helium–neon gas laser and a maximum visible radiant power of one milliwatt.

The accuracy of lasers, for levelling, is similar to that of levels, but although the beam is narrow it does diverge slightly and at long distances, or in strong sunlight, a sensor is used to detect the beam. A sensor is necessary, of course, on models which do not provide a visible beam. The sensor may be powered or manually controlled to move up or down a rod until the correct position is reached. Like a level, the instrument must be checked from time to time to check that the beam is on the correct alignment. Again, like other instrumental sight lines, the beam is susceptible to the effects of curvature and refraction.

Like any other bright light, lasers can cause permanent damage to the eyes and the light source and any reflection or refraction of it should not be stared at or viewed through a theodolite or level. Wherever possible, lasers should be set up well above or below eye height. Two new documents on the use of lasers have been published recently, BS 4803: 1983: *Radiation Safety of Laser Products and Systems* and the *RICS Laser Safety Code*. The British Standard recommends that a Laser Safety Officer should be on site where lasers with a radiant power greater than 1 mW are in use.

17.4.3   Common setting out tasks

The tasks involved in height control may be reduced to establishing a point or mark at some given height, establishing two or more such points to define a horizontal or inclined line or plane and establishing a series of points in a curve in a vertical plane, this latter then being known as a *vertical curve*.

17.4.3.1   *Establishing a point at a specified height*

The plan location of the required peg or mark must be fixed first and in this case, it will be assumed that a peg is to be placed as a height marker. A peg must be driven at the required position, ensuring that the top of the peg is above the required height, then the level of the top of the peg determined by careful levelling from a TBM. When the peg top level is known, the difference between peg level and required level may be scaled down the side of the peg and a horizontal mark drawn across the peg to indicate the desired height reference. As a check, the levelling staff zero should be placed against the drawn line and the height of the mark determined again by levelling and calculating its reduced level.

17.4.3.2   *Establishing horizontal or sloping lines or planes*

These tasks are carried out by placing appropriate pegs as described in Section 17.4.3.1, according to the demands of the specific task.

17.4.3.3   *Vertical curves*

The longitudinal profile of a road consists of straight lines joined by curves, similar to its plan, as in *Figure 17.11(a)*. Vertical curves between straight

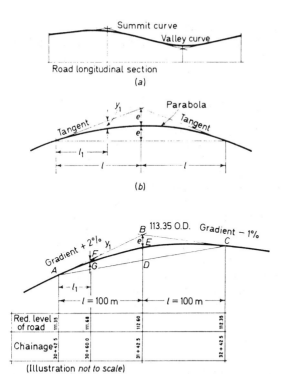

Figure 17.11

tangents are not, however, formed by circular arcs but by parabolic curves. *Figure 17.11(b)* shows the important features of such a parabola: e is the maximum (central) ordinate of the curve from the intersection point, l is the half-span or half-length, l_1 any distance from one end of the curve and y_1 the ordinate from the tangent to the curve at the end of distance l_1. The equation to find any ordinate such as y_1 is $y_1/l_1^2 = e/l^2$ or

$$y_1 = e(l_1/l)^2.$$

In practice, the two road straights or tangents seldom slope uniformly on either side of the intersection, and it is conventional to measure l and l_1 horizontally and e and y_1 vertically, the errors being negligible. The first decision required is as to the *length* of the curve, $2l$, but it is not intended to cover the design of curves here. Typically, in a small job, the curve length will be specified, and the task is merely to set out the appropriate distances and levels in the field. (Normal setting out measurements.) The method is best shown by a simple example.

In *Figure 17.11(c)* the intersection of two road straights drawn on a longitudinal section is at point B, chainage 31 + 42.5, reduced level 113.35 metres AOD. The gradients of the straights (tangents) are AB + 2%, BC − 1%. The curve is to have a length of 200 m, i.e. $l = 100$ m. The reduced level is required for each even chainage point along the curve. (The first point only, at chainage 30 + 60.0, will be calculated.)

The levels of A and C are readily calculated as 111.35 and 112.35 AOD respectively.

Level of D = (112.35 + 111.35)/2 = 111.85
Level of E = (113.35 + 111.85)/2 = 112.60
BE = ED = e = 112.60 − 111.85 = 0.75
l_1 = 3060.0 − 3042.5 = 17.5
y_1 = 0.75(17.5/100)^2 = 0.023
Level of F = 111.35 + (2 × 17.5/100) = 111.70
Level of curve at G = 111.70 − 0.02 = 111.68 AOD

Similar calculations may be made for chainages 30 + 80.0, 31 + 00.0, etc., a tabular form being suitable for calculation and checking purposes.

17.5   Vertical alignment control

The control of vertical alignment in construction works has two distinct aspects. These are first, the vertical transfer of control points or lines to higher or lower levels and second, the provision of vertical control lines and the checking of verticality of construction elements.

17.5.1   Vertical transfers of control points or lines

The methods for establishing a reference frame at ground floor level have been outlined already. For the plan control of higher floors, or basements, etc., it is necessary either to reconstruct the original reference frame at the new levels or to transfer fixed points of the ground frame to the new levels and use these to construct new reference frames.

As usual, the accuracy required for vertical position transfer depends upon the particular job specification. For tall buildings, a typical requirement is that the error in position of the reference lines at tenth floor level should not exceed ±2.5 mm.

BS 5964: 1980 specifies a permissible deviation of ±2/L mm where L is the vertical displacement in metres between the main and transferred point, with a permissible deviation of ±4 mm where L is less than 5 m.

The *plumb-bob and string* is widely used on small buildings and within individual stories of large buildings, but precision is difficult for large height differences. BRE Digest 234, Feb. 1980, *Accuracy in Setting-Out*, suggests a maximum range of 3 m and probable accuracy of 5 mm at this range. Heavy plumb-bobs should be used, of 3 to 6 kg.

Vertical transfer is often carried out by theodolite intersection. In this method, a ground point is observed from two directions in plan (preferably at right angles) then the point may be re-located at high level by elevating the theodolite telescope and sight line. This is similar in principle to the method of prolonging a straight line. The method is easier and faster if two theodolites are used.

The principal method of vertical transfer for the control of large buildings is optical plumbing, where an instrument sight line is directed vertically upwards from ground control points, either using a specialist instrument or a theodolite with special attachments. Variations of this method make use of the laser to provide a visible vertical line.

17.5.2   Vertical transfer by optical plumbing

Although there are certain tasks for which the suspended string or wire plummet is the best method of vertical position transfer, optical plumbing is generally the most effective method in the construction of tall buildings.

The normal technique is to locate control points at ground level then set an instrument exactly over these and plumb upwards to locate new reference points on each floor in turn. On each floor, the new points are used to set out reference lines for the positioning of construction elements on that floor. The points are located at high level by using suitable aiming targets according to the particular circumstances. For plumbing in open-frame structures, or outside buildings, aiming targets on offset brackets may be used. When plumbing inside buildings which have the floors laid or constructed as the building rises, four or more (as necessary) holes, about 200 mm square, are left open in each floor in suitable positions near the corners of the floor plan. When a floor is formed, aiming targets or graduated scales are placed over each hole and accurately positioned for the floor reference frame setting-out. When work is complete on the floor, the targets are removed to allow clear sighting from the instrument at ground level up to the targets on the next floor.

Although some instruments have ranges of 100 or 200 m, the practical limit for optical plumbing is generally taken as about 10 stories or about 30 m. All floors up to the tenth are, therefore, fixed by sighting from ground level, then the instrument is set up again at the tenth floor level and sights are made up to the twentieth floor and so on.

Targets for sighting are often best designed for the individual job, depending upon the instrument positions, the structure, the sight length and the instrument to be used. A target may be as simple as a board with the plumb-point pencilled on the underside, then a nail driven through the mark to provide an instrument station on the upper surface, or alternatively graduated transparent sheets or scale rules.

With all optical plumbing instruments, the vertical observation should be made from four distinct orientations of the instrument in plan, at 90° to one another, then the mean position of the four target points defined should be free from instrument error.

17.5.3   Optical plumbing instruments

Optical plumbing instruments all rely on either a bubble tube or an automatic levelling compensator, then the horizontal sight line is deviated through 90° to provide a vertical sight line.

17.5.3.1   Theodolite telescope roof plummet

This is a small horizontal telescope with a 90° prism built in. It is attached to the top of a theodolite telescope and when the main telescope is exactly horizontal it provides a vertical line of sight. The particular application of this attachment is in locating a theodolite under an overhead mark, or in fixing an overhead mark above a theodolite station. The Wild version, used on T1, T16 and T2 theodolites, has range of 10 m, the accuracy claimed is 1:5000.

17.5.3.2 Small independent optical plummets

These consist of a small horizontal telescope fitted with a pentaprism to provide a vertical sight line, supported by a theodolite tribrach and tripod. Levelling up is by reference to bubble tubes. Relatively short range, accuracy possibly 1:10 000.

17.5.3.3 Theodolite with diagonal eyepieces

Most instrument manufacturers produce *long diagonal eyepieces* for their theodolites and when these are fitted it is possible to observe objects at large angles of elevation (up to the zenith) and also to read the circles. Provided the theodolite is in good adjustment, then when the vertical circle is set to 0° (or 90°, as appropriate) the sight line is directed to the zenith tracing a vertical line through the instrument centre. The theodolite's own optical plummet may be used for centring over the ground mark, in the usual way.

BRE Digest 234 suggests that a 20″ theodolite used in this way (and remembering to take four sets of observations at 90° in plan) will give a probable accuracy of ±5 mm at a maximum range of 50 m, or 1:10 000. Using a 1″ theodolite, ±5 mm at 100 m, or 1:20 000.

17.5.3.4 Objective pentaprism

This is a prism which is attached to the objective end of an automatic level's telescope, deflecting the line of sight through 90°. Wild claim that it converts their NA2 automatic level into a high precision optical plummet for upward or downward plumbing, with an accuracy of 1:100 000. The pentaprism may also be used with a Wild theodolite, when 1:70 000 is claimed.

Note that with this instrument, it must again be used in four directions in plan and opposite pairs of indicated points must be connected so that the intersection of the lines defines the plumb point. (When a pentaprism is fitted on a theodolite telescope, the telescope must be sighted horizontally for plumbing and thus with both instruments the prism provides a vertical line which is offset from the plan centre of the instrument.)

17.5.3.5 Independent precision plummets

These are similar to the instrument described in Section 17.5.3.2, but with a high power telescope and a built-in pentaprism. Verticality may be controlled either by bubble tubes or an automatic compensator unit like an automatic level. Some versions have two telescopes, sighting respectively up and down, others a single telescope sighting one way only. Manufacturers claim accuracies between 1:50 000 and 1:200 000, with ranges of 100 to 200 m. BRE Digest 234 suggests a probable accuracy of ±5 mm at a maximum range of 100 m.

17.5.4 Vertical transfer by laser

Alignment lasers which project a *vertical reference line* may be used for plumbing, accuracy being checked by rotating the laser instrument about its vertical axis in the same way as for optical plummets. It should be noted, however, that a laser costs nearly twice as much as a precise optical plummet and about one-and-a-half times as much as a suitable theodolite.

A laser beam projected through a precise optical plummet was used to control the verticality in setting out Australia's tallest structure, the 300 m high Centrepoint Tower in Sydney, primarily a 250 m high 7 m diameter steel shaft.

Laser attachments may be obtained for theodolites, these using a beam splitter so that the laser beam is projected along the collimation line while the observer can sight along the collimation line through the eyepiece in the usual way. Such devices may be used to sight at any inclination.

17.5.5   Vertical control lines and checking verticals

17.5.5.1   Vertical control lines

These lines are often required so that walls, columns, lift guide rails, etc., may be erected properly vertical. Traditionally the best example is the *suspended plumb-bob*, providing a visible, physical line from which craftsmen may measure offset distances themselves as required. When the plumb-bob has settled and stopped swinging and the line is correctly positioned, the lower end of the wire or string may be made fast and tightened by turnbuckle. Today the plummet is being replaced by the laser, either an *alignment laser* projecting a vertical line or a *rotating laser* in a *vertical plane* being used to identify the position of objects such as internal walls.

17.5.5.2   Checking verticals

The checking of verticality of columns, walls, stanchions, etc., may be carried out with the plumb-bob and string as in the previous section.

The theodolite or a rotating laser set in a vertical plane may be used in steel erection to check stanchions for both plumb and alignment. The method is to set out a line parallel to the centre-line of the stanchions and offset some distance from it and set up the instrument on the line but well back from the nearest stanchion. If the instrument is aligned accurately along the offset line, any point on any stanchion may be checked by an offset measurement from the collimation plane or the rotating beam plane. To check plumb, the procedure must be repeated with other offset lines at 90° in plan.

Optical plummets and alignment lasers may also be used for checking verticals.

17.6   Excavation control

The typical tasks in excavation or earthworks are (1) excavation or fill over an area, to a new level, (2) excavation for cutting and filling embankments for roads, etc. and (3) excavation for trenches and pipelaying. The general control for heights is provided by the usual site benchmarks, located well away from areas of activity and plant movement.

17.6.1   Area excavation or fill

Traditionally, height control pegs should be placed over the area concerned, their tops carefully levelled from site benchmarks. To *fill* over an area, pegs are best driven until their heads are at the required new level. Where this is impracticable, pegs should be marked with the new level, or the distance to be measured vertically to reach the new level. (See *Figure 17.12* and note that in (*d*) the top of the peg is 0.5 m above the new level required – ambiguity must be avoided.)

For *excavating* over an area, pegs may be placed similarly and marked to show how high the peg top is above the required level, see *Figure 17.12(e)*.

To provide site operative control or earthworks, *sight rails* or *profiles* may be set up as in *Figure 17.12(c)* or (*f*). The sight rails should be

EXCAVATION CONTROL

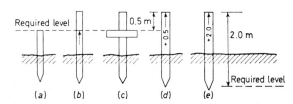

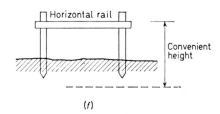

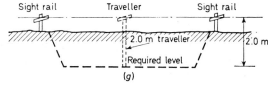

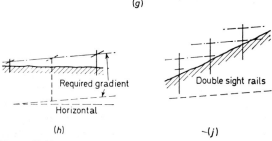

Figure 17.12

constructed at a convenient viewing height of 1 to 1.5 m above existing ground level, such that a line of sight between the tops of any pair is parallel to the required level and preferably at some multiple of 0.1 m. Site operatives use these rails in conjunction with a *traveller* or *boning rod* to control excavation or fill by sighting from one sight rail to the next. When the traveller rests on the surface of the earth and its top is in line with the line of sight between the rails, the ground is at the correct level, see *Figure 17.12(g)*.

Gradients may be established in the same way, with the sight rails set so as to be parallel to the plane of the slope and at some convenient height above it, *Figure 17.12(h)*. *Double sight rails* may be required on steep slopes, *Figure 17.12(j)*.

Earthworks are carried out to the levels required between lines of pegs or sight rails, leaving these standing on mounds and finally the mounds are cut away to level up the remainder.

An alternative approach is to use a rotating laser and to dispense with sight rails and travellers, although markers are necessary to indicate the extent of the area to be regraded. The laser is set up clear of the working area, at such a position and height that the beam will not be at the eye level of the site operatives and others adjacent to the site and such that the rotating beam is at some convenient constant height above the finished plane surface. *Sensor units* may be attached to the earth moving vehicles at the appropriate height, with the detectors outside and the receivers inside the cab.

17.6.2   Cut and fill for roads, etc.

These are essentially controlled for height in the same way as general areas, the difference lying in the formation of the side slopes. The position of the limits of side slopes is indicated by *slope stakes* or *batter pegs*. These are often offset at a standard distance beyond the outer limit of the slope, for example one metre, since if placed on the exact alignment they may be disturbed by the site operatives. The position of these pegs is obtained either by scaling off from the working drawings or by a combination of scaling and calculation to obtain compatibility between the distance and the height difference from the centre-line to the outer limit of the side slope, maintaining the correct side slope gradient. Again, sight rails, also known as *batter boards*, are used to control the angle of slope and may be constructed at a convenient height above the slope, or on the slope, or often on a prolongation of the line of slope, see *Figure 17.13*.

17.6.3   Trench excavation

The traditional method of control is by an application of *sight rails* and *traveller*. In this case, either a *two-post sight rail* is set across the trench line or, if mechanical excavation is to be used, a *T-shaped sight rail* is set to one side of the trench. The alignment of the trench and the pipe centre is controlled either by marks on the two-post sight rails or by offset pegs.

The *traveller* may be used to set out and check the following – the depth of trench, the thickness of the bed upon which the pipe is to rest and the

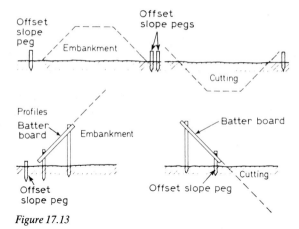

Figure 17.13

level of the inside bottom surface of the pipe, known as the invert level. This is achieved by having the traveller made to be correct at the top of the bed, an extension piece being added for the initial excavation, and an angle bracket fitted for the invert level, see *Figure 17.14*.

An alternative approach is to use an alignment laser to control both the formation of the trench and the laying of the pipe. Initially the sensor may be in the mechanical excavator and, later, the beam may be directed down the centre of the pipe as it is laid.

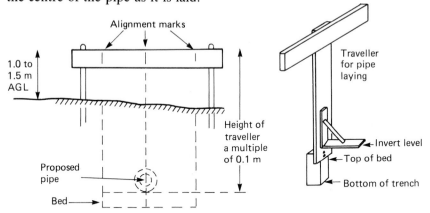

Figure 17.14

17.7   Microcomputer applications

The MICROSURVEY system can provide assistance with most of the tasks covered by this chapter.

17.7.1   Trigonometry, co-ordinates, unit conversions, areas, etc.

The Trigonometry program may be used for triangle solution, co-ordinate problems of intersection, etc., conversions between different units of measurement and area and volume calculations.

17.7.2   Setting out by bearing and distance

The Set Out program may be used to compute the horizontal angle and distance for setting out a ground point from an instrument station, given that the rectangular co-ordinates of the ground point, of the instrument station and of either 1 or 2 Reference Objects are known.

The horizontal angle calculated is the angle between the lines from the instrument station to the RO and from the station to the required location of the co-ordinated setting out point. Where two ROs are used, two angles and distances are calculated, this may be desirable as a check in some circumstances.

```
Leicester  Polytechnic

SETTING  OUT  DETAIL

Setting out detail printout from CBM micro program SET  OUT
_____

     Copy 1     Dated 7th August 1984 (Data held on file se-setoutdemo)

     Job name :  Program test - inversion of LRI data from DI3 survey

     Client   :  R E Paul

     Surveyor : W.S.Whyte on 9th May 1980
Setting out Station and Point Detail
_____

Station   1 (401/lri  )      Co-ordinates      Bearings

                  Stn.    696.750   762.950
                  R.O.    794.000   591.900   150 22 47 (ro401/lri )

Points set out from Station 1
_____

                                        Horizontal Angles
        E          N        H.Dist.     from R.O.  from check R.O.

Point 1  (407st)
    793.999    591.901     196.762      00 00 00
Point 2  (a    )
    679.205    752.434      20.455      88 40 59
Point 3  (b    )
    659.652    742.798      42.218      91 06 33
Point 4  (c    )
    644.494    740.016      57.067      95 55 29
Point 5  (d    )
    643.853    740.438      57.488      96 34 00
Point 6  (e    )
    641.657    739.683      59.805      96 43 29
Point 7  (f    )
    578.941    713.902       *
Point 8  (g
```

Figure 17.15

The program can handle up to 25 points from each of 4 stations. Where more than 25 points are to be located from a station, then the station details may be repeated for a second group of up to 25 points, and so on. The usual disk file data storage, amendment, printout and display routines are available, but printout may be in abbreviated or full format. The short print is suited to site use.

Figure 17.15 shows part of a full printout for setting out. The given co-ordinates of the points to be set out are shown on the left, the calculated distances and angles from the RO(s) to be used in the field are on the right.

```
Leicester  Polytechnic

HORIZONTAL  CURVE
```

Circular curve data printout from CBM micro program HORIZONTAL

Copy 1   Dated 7th August 1984 (Data held on file h-circdemo)

Job name : Advanced Engineering Survey Ex. 5.10

Client   : Commodore

Surveyor : W S Whyte

Horizontal Circular Curve

Curve Ex. 5.10   (right)

Tangent alignment co-ordinates specified

Entry tangent			Exit tangent		
	E	N		E	N
Point A	2675.454	3748.621	Point C	2965.131	5087.163
Point B	2845.736	4972.814	Point D	3047.162	3796.734

Curve specification

Radius	Degree	BC Chainage	Pt.A Chainage
400.000	14 19 26	29+14.034	50+00.000

Through chainage - standard chord length of 20

Computed data

Tangent length	3952.951	Long chord length	795.935
Curve length	1175.960	Rise of long chord	359.730
Apex distance (ED)	3573.138	Deflection angle	168 26 38

	Chainages	E	N
Beginning of Curve	29+14.034	2388.068	1682.546
Crown Point of Curve	35+02.013	2799.196	2027.159
End of Curve	40+89.993	3183.448	1652.814
Intersection Point	68+66.985	2932.670	5597.803
Centre of Circle		2784.254	1627.438

First sub-chord   5.966   58 Standard chords Final Sub-chord   9.993

Tangential angle for standard chord -   1 25 57   5156.6 secs

Continued -

Figure 17.16

Page 2 Curve right Ex. 5.10

Curve point details

					Tangential Angles	
Peg	Chainage	Chord	E	N	At BC	At EC
BC	29+14.034		2388.068	1682.546	00 00 00	84 13 19
		5.966				(275 46 41)
1	29+20.000		2388.934	1688.449	00 25 38	83 47 41
		20.000				(276 12 19)
2	29+40.000		2392.478	1708.131	1 51 35	82 21 44
		20.000				(277 38 16)
3	29+60.000		2397.000	1727.611	3 17 32	80 55 47
		20.000				(279 04 13)
4	29+80.000		2402.491	1746.840	4 43 28	79 29 51
		20.000				(280 30 09)
5	30+00.000		2408.936	1765.771	6 09 25	78 03 54
		20.000				(281 56 06)
6	30+20.000		2416.318	1784.356	7 35 21	76 37 57
		20.000				(283 22 03)
7	30+40.000		2424.621	1802.549	9 01 18	75 12 01
		20.000				(284 47 59)
8	30+60.000		2433.822	1820.305	10 27 15	73 46 04
		20.000				(286 13 56)
9	30+80.000		2443.900	1837 570	.. -	
		20.000				
10	31+00.000					

Figure 17.17

17.7.3   Line reduction and levelling

As outlined earlier, site-measured distances may be corrected using the Lines program and site levelling may be reduced and adjusted with the Levels program.

17.7.4   Horizontal curves

The Horizontal program may be used to compute the rectangular co-ordinates of points on a circular horizontal curve, together with the tangential deflection angles to these points from both ends of the curve. Given the appropriate co-ordinate data to define the tangent alignments, together with the chainage of the one point on the entry tangent, the curve may be designed on the basis of required radius, or degree of curve, or start point chainage. Curve points may be on 'through chainage' or equal chords, with any specified chord length, to a maximum of 200 points. Where the tangents are not defined by co-ordinates, then of course arbitrary co-ordinates may be allotted for the purposes of the curve design.

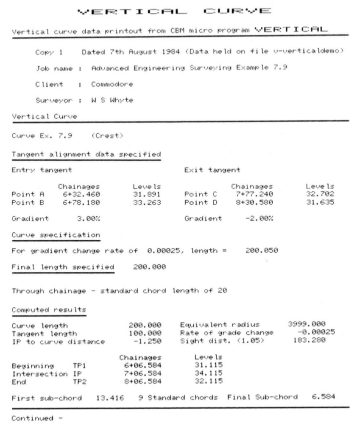

Figure 17.18

340   Setting out

Page 2   Curve Ex. 7.9   (Crest)

Curve point details

Peg	Chainage	Chord	Tan. Level	Offset	Curve Level	Offset	Tan. Level
TP1	6+06.584		31.115	0.000	31.115	-5.001	36.116
		13.416					
1	6+20.000		31.517	-0.023	31.495	-4.353	35.847
		20.000					
2	6+40.000		32.117	-0.140	31.978	-3.470	35.447
		20.000					
3	6+60.000		32.717	-0.357	32.361	-2.687	35.047
		20.000					
4	6+80.000		33.318	-0.674	32.644	-2.003	34.647
		20.000					
5	7+00.000		33.918	-1.091	32.827	-1.420	34.247
		20.000					
6	7+20.000		34.518	-1.608	32.910	-0.937	33.847
		20.000					
7	7+40.000		35.118	-2.226	32.893	-0.554	33.447
		20.000					
8	7+60.000		35.718	-2.943	32.776	-0.271	33.047
		20.000					
9	7+80.000		36.318	-3.760	32.558	-0.088	32.647
		20.000					
10	8+00.000		36.919	-4.677	32.241	-0.005	32.247
		6.584					
TP2	8+06.584		37.116	-5.001	32.115	0.000	32.115

MICROSURVEY Software by Construction Measurement Systems Ltd
Copyright (C) W.S.Whyte 1983 Tel: (0537 58) 283
8000/2.1/8050/28 - for Leicester Polytechnic - 12.1.84

Figure 17.19

Figure 17.16 shows the first page of a curve printout, containing a summary of all data and design information. *Figure 17.17* shows part of the computed curve point details, including both rectangular co-ords for all points and the tangential deflection angles from both ends. Disk file data storage, amendment, printout and display routines are provided and, in addition, a file of co-ordinates written by this program may be read by the Transform and Co-ordlister programs. Co-ordinates computed in this program may be entered in the Set Out program for setting out the points on site by bearing and distance.

Transition and Cubic Spline programs are available in later versions of MICROSURVEY.

17.7.5   Vertical curves

The Vertical program may be used to compute the chainages and reduced levels of points on a vertical highway curve, together with their offsets from both tangents to the curve. Given the appropriate chainages and levels to define the tangent alignments, the curve may be designed on all the usual bases. Curve points may be on equal chords or on 'through chainage', with any specified chord length, to a maximum of 100 points.

Figures 17.18 and *17.19* show an example vertical curve printout. Disk file data storage, amendment, printout and display routines are included as usual.

17.7.6   Co-ordinate transform, sort and list

Co-ordinate lists may be transformed, listed and sorted using the Transform and Co-ordlister programs, as outlined earlier.

Bibliography

Allen, A. L., Hollwey, J. R. and Maynes, J. H. B., *Practical Field Surveying and Computations*, Heinemann, (1968).

Badekas, J, (ed.), *Photographic Surveys of Monuments and Sites*, Elsevier, (1975).

Bowyer, J., *Guide to Domestic Building Surveys*, Architectural Press, (1979).

Brandes, D. 'Sources for relief representation techniques', *The Cartographic Journal*, **21,** No. 2, (1983).

Brighty, S. G., *Setting Out – a Guide for Site Engineers*, Granada, (1975).

British Standards Institution, BS 1377, *Methods of Testing Soil for Civil Engineering Purposes*, (1975).

British Standards Institution, PD 6479, *Recommendations for Symbols and other Graphic Conventions for Building Production Drawings*, (1976).

British Standards Institution, BS 5964, *Methods for Setting Out and Measurement of Buildings: Permissible Measuring Deviations* (1980).

British Standards Institution, BS 5930, *Code of Practice for Site Investigations*, (1981).

British Standards Institution, BS 4803, *Radiation Safety of Laser Products and Systems*, (1983).

Building Research Establishment, CP 15, *Accuracy Achieved in Setting Out with Theodolite and Surveyor's Level on Building Sites*, HMSO, (1977).

Building Research Establishment, Digest 202, *Site Use of the Theodolite and Surveyor's Level*, HMSO, (1977).

Building Research Establishment, Digest 234, *Accuracy in Setting Out*, HMSO, (1980).

Building Research Establishment. Digests 63; 64; 67, *Soils and Foundations*, HMSO, (1980).

Collins, B. J. and Madge, B., 'Photo-radiation – a new method of monitoring beach movement', *Chartered Land Surveyor/Chartered Mineral Surveyor*, **3,** No. 1, (1981).

Eldridge. H. J., *Common Defects in Buildings*, HMSO, (1976).

Feihl, O., 'Surveying instrument in architecture and archeology', *Bulletin Kern*, **33,** (1982).

Fraser, S., 'Developments at the Ordnance Survey since 1981', *Cartographic Journal*, **21**, No. 1, (1984).

Keates, J. S., *Understanding Maps*, Longman, (1982).

McCullough, W. R., 'The measurement of distance using light', paper presented to the 15th Survey Congress of the Institution of Surveyors of Australia, (1972).

Ordnance Survey of Great Britain, *Projection Tables for the Transverse Mercator Projection of Great Britain*, HMSO, (1950).

Ordnance Survey of Great Britain, *Some Facts and Figures Relating to the Transverse Mercator Projection and the National Grid*, Old series leaflet No. 37, (1975).

Ordnance Survey of Great Britain, *Ordnance Survey Maps: a Descriptive Manual*, (1975).

Reekie, F., *Draughtsmanship*, 3rd. edn, Arnold, (1976).

Schofield, W., *Engineering Surveying 1*, Butterworths, (1984).

Shepherd, F. A., *Surveying Problems and Solutions*, Arnold, (1968).

Shepherd, F. A., *Advanced Engineering Surveying*, Arnold, (1981).

Smith, J. R., *AGA Geodimeter*, 1947–1978.

Waalenijin, A., 'Hydrostatic Levelling in the Netherlands', *Survey Review*, Nos 131 and 132, (1964).

Wadley, T. L., 'The Tellurometer System of distance measurement', *Empire Survey Review*, Nos 105 and 106, (1957).

Wadley, T. L., 'Electronic principles of the Tellurometer', *Transactions of the South African Institute of Electrical Engineers*, (1958).

Whyte, W. S., *Revision Notes on Plane Surveying*, Butterworths, (1971).

Whyte, W. S. and Paul, R. E., *Site Surveying and Levelling 2*, Butterworths, (1982).

Wilson, R. J. P., *Land Surveying*, Macdonald and Evans.

Index

343